Fundamentals of Building Performance Simulation

Fundamentals of Building Performance Simulation pares the theory and practice of a multi-disciplinary field to the essentials for classroom learning and real-world applications. Authored by a veteran educator and researcher, this textbook equips graduate students and emerging and established professionals in engineering and architecture to predict and optimize buildings' energy use. It employs an innovative pedagogical approach, introducing new concepts and skills through previously mastered ones and deepening understanding of familiar themes by means of new material.

Covering topics from indoor airflow to the effects of the weather, the book's 19 chapters empower learners to:

- Understand the models and assumptions underlying popular BPS tools
- Compare models, simulations, and modelling tools and make appropriate selections
- Recognize the effects of modelling choices and input data on simulation predictions
- And more.

Each subject is introduced without reference to particular modelling tools, while practice problems at the end of each chapter provide hands-on experience with the tools of the reader's choice. Curated reading lists orient beginners in a vast, cross-disciplinary literature, and the critical thinking skills stressed throughout prepare them to make contributions of their own.

Fundamentals of Building Performance Simulation provides a much-needed resource for new and aspiring members of the building science community.

Ian Beausoleil-Morrison is a Professor in the Faculty of Engineering and Design at Carleton University where he is the Canada Research Professor in Innovative Energy Systems for Residential Buildings. His research interests include solar housing, seasonal thermal storage, and building performance simulation. Prior to joining Carleton University in 2007, he worked for 16 years at CanmetENERGY (a Canadian government laboratory) where he led a team of researchers developing building simulation models and tools to support industry and government programmes.

Professor Beausoleil-Morrison was President of IBPSA from 2010 to 2015, Vice-President from 2006 to 2010, and founded IBPSA-Canada and initiated the eSim conference series in 2001. He also cofounded and is the Co-Editor of IBPSA's *Journal of Building Performance Simulation*, and is a Fellow of IBPSA.

Advance Praise for Fundamentals of Building Performance Simulation

"The effective application of building performance simulation is governed by an important caveat: that users possess an understanding of the underlying physics and the related design parameters that influence performance outcomes. This tour de force contribution from a leading proponent of the technology precisely hits the spot and will surely raise the standard in simulation applications worldwide." –**Joseph Andrew Clarke, University of Strathclyde**

"*Fundamentals of Building Performance Simulation* is set to be the authority on the subject, making it essential reading for all students and junior practicing engineers working in the area of building energy and performance" –**Malcom Cook, Loughborough University**

"For many years I've been looking for a book that was explicitly constructed to teach building performance simulation. Professor Beausoleil-Morrison has taken his teaching approach—insightful review of theory, supplemented by required reading, then hands-on simulation exercises and results autopsy, and review—and turned it into a textbook. Anyone with a basic understanding of building physics and thermodynamics will find this book robust, without depending on any one simulation program. Yet it directs readers how to test the underlying physics and concepts using their preferred simulation tools. If I were looking for a book to use in teaching building performance simulation, this would be my textbook." **–Dru Crawley, Bentley Systems, Inc.**

"For learning from a leading expert and teacher in using building performance simulation in a responsible manner, Ian Beausoleil-Morrison presents the theoretical and practical fundamentals in this excellent textbook." **–Jan Hensen, Eindhoven University of Technology**

"The art of building performance simulation is to capture all important physical processes in mathematical equations and make them converge to the right solution. Ian Beausoleil-Morrison has mastered better than anyone how to do this, how to explain this and how to teach this, turning this book into a milestone for anyone working towards a sustainable built environment." **–Lieve Helsen, University of Leuven**

"Many countries are committed to mitigating climate change though energy efficient buildings with integrated renewable energy generation. Integrated building design is paramount to highly efficient buildings. Building simulation is fundamental in this process. Prof. Beausoleil-Morrison brings us a fantastic textbook, based on his experiential teaching method, that helps us to develop the critical view so necessary in a building simulation professional. It will become a 'must have' for all universities with courses in this area." **–Roberto Lamberts, Laboratory for Energy Efficiency in Buildings, Federal University of Santa Catarina**

"Did you ever wonder what the strengths and limitations of the models in building performance simulators are? This book explains rigorously, yet approachably, the major models that can be found in today's building performance simulators. This is recommended reading for anyone who needs to be competent in building performance simulations." **–Michael Wetter, Lawrence Berkeley National Laboratory**

"It's an ideal textbook for teaching and self-study. Professor Beausoleil-Morrison not only gives a complete introduction to the basic principles and tools but also provides all the necessary teaching materials. I would be happy to recommend it to my colleagues who are teaching BPS and students who are learning it." **–Yingxin Zhu, Tsinghua University**

Fundamentals of Building Performance Simulation

Ian Beausoleil-Morrison
https://orcid.org/0000-0001-7123-841X

Routledge
Taylor & Francis Group

NEW YORK AND LONDON

First published 2021
by Routledge
52 Vanderbilt Avenue, New York, NY 10017

and by Routledge
2 Park Square, Milton Park, Abingdon, Oxon, OX14 4RN

Routledge is an imprint of the Taylor & Francis Group, an informa business

Library of Congress Cataloging-in-Publication Data
Names: Beausoleil-Morrison, Ian, author.
Title: Fundamentals of building performance simulation / Ian Beausoleil-Morrison.
Description: New York : Routledge, 2020. | Includes bibliographical references and index.
Identifiers: LCCN 2020029054 (print) | LCCN 2020029055 (ebook) | ISBN 9780367518066
(paperback) | ISBN 9780367518059 (hardback) | ISBN 9781003055273 (ebook)
Subjects: LCSH: Buildings–Performance–Simulation methods.
Classification: LCC TH453 .B44 2020 (print) | LCC TH453 (ebook) | DDC 720/.472011–dc23
LC record available at https://lccn.loc.gov/2020029054
LC ebook record available at https://lccn.loc.gov/2020029055

ISBN: 978-0-367-51805-9 (hbk)
ISBN: 978-0-367-51806-6 (pbk)
ISBN: 978-1-003-05527-3 (ebk)

Typeset in LATEX TX fonts by Ian Beausoleil-Morrison

Publisher's Note: This book has been prepared from camera-ready copy provided by the author

Visit the companion website: www.Routledge.com/cw/beausoleilmorrison

À Rachel, Amiel, et Rhiannon.
Merci de votre appui sans réserve.
Je vous aime.

Contents

PART IV **Building envelope**

Part V **HVAC**

Part VI **Finale**

Foreword

In the face of overwhelming evidence that human activity is adversely influencing the ecological balance of the earth, and the associated fact that the built environment and transport account for more than half of the world's energy consumption and emissions, it is incumbent on those designing our shared future environments to do so with great care (Brundtland Report, 1987).

According to the Brundtland definition, we, the current generation, should "meet our needs without compromising the ability of future generations to meet their own needs." But in a rapidly evolving world these needs are constantly changing and increasing in complexity. This has to be taken into account when making predictions and decisions related to an uncertain future while striving to balance environmental concerns with the wider needs of everyone on the planet—all of which will affect how we design our buildings and cities. Designers need to be better equipped to deal with this uncertainty.

Computer simulation of buildings in the early days was an activity confined to architectural, building physics and engineering research labs. However, for a variety of reasons, from the 1970s onwards, interest in the potential of 'computer simulation' to improve design began to emerge from within progressive design practices—despite no knowledge of tools or how to use them. While the opportunities in terms of quantifying and refining the potential energy and environmental benefits were self-evident, the professions had to find a way of overcoming many barriers, such as those caused by the time pressures of the design process, a lack of trust in new methods and the cost and time investments required to adopt these new methods.

Over time, as computing power has increased and designers have almost universally adopted a digital approach to design, many of the previously perceived barriers have ceased to exist. We now have access to high-power computers and building performance simulation tools with seemingly user-proof interfaces that can undertake simulations quickly. But this brings with it the need for better understanding of the consequences of input decisions: simplified interfaces can make things that are really complex seem simple, so the need for user understanding is paramount. Whereas in the past modellers had to seek out advice on input parameters, many of the blanks are now filled in with default values, allowing the modeller to rapidly develop skills to operate tools without necessarily understanding the consequences of their (sometimes random) choices.

It was while working as a researcher at the University of Strathclyde's Energy Systems Research Unit (ESRU) in the late 1990s that I first met the author, Ian Beausoleil-Morrison. At that time our research interests were very different—his PhD research focused on advancing whole-building simulation development through the conflation of dynamic simulation with computational fluid dynamics, and my interests were more in practical application. However, over time our interests have converged.

While my work has remained focused on exploring the potential for embedding building performance simulation in design practice, Ian's research interests have increasingly been focused on equipping users, and students in particular, with the skills and knowledge required to use simulation responsibly.

To this end, Ian has collaborated with others worldwide and published extensively on the effectiveness of teaching of building performance simulation through a complete and continuous learning spiral, enshrining experiential (doing) rather than passive (listening) teaching methods to guide students, researchers and those entering the design professions on gaining the skills, knowledge and understanding required to interpret, scrutinize and verify simulation predictions. The approach is predicated on the application of tools from the start to reinforce an appreciation of what? how? and why? including the potential impact of using tool default methods and data, the myriad sources of uncertainty and the roles of interrogation and analysis, with one topic leading seamlessly into another.

In recent years there have been a number of books published on the subject of building performance simulation—some on the underlying fundamentals, some on the application possibilities and some on use in practice, for audiences ranging from undergraduates to researchers to tool developers to practitioners. However, I have been concerned for a long time that nothing existed that supported a computational approach to building appraisal while providing the modeller with guidance on the inherent risks. For this reason, I am delighted that this book has been written, it is a much needed and overdue link between theory and practice. Furthermore, I consider Ian to be ideally placed to write this book—he has appreciated the issues involved from all angles: as a tool developer, from the policy delivery side, supporting practioners and as an educator.

As students, researchers, developers and practitioners of the built environment, we all have a responsibility to make sure that we use simulation appropriately, and that we use it to give good advice, but to date there has been little guidance available beyond the lecture hall or seminar. This book takes students of building performance simulation through a programme that subtly but clearly highlights and links the complexities and the importance of good decision making by purposely providing readers with simple task-based modelling exercises that incrementally grow in complexity, allowing the student to both develop their skills and understand the consequences of good and bad decision-making.

The book highlights that in addition to an understanding of the physics of building design and performance, it is also important to appreciate what is critical all of the time and what matters only in particular circumstances. This is where the art comes face to face with the science. While the book provides detailed technical explanations for the phenomena described, this is done without getting in the way of learning through practical application: the two run parallel but are always equally in sight, striking a balance between understanding complex thermophysical interactions while learning to positively inform performance outcomes.

The extended reading lists and task-based exercises support confidence-building through dealing with the abstraction of complex realities. If they stay the course, by the end, students will know what they know and appreciate their limitations and whether or not it matters. A key point is that the author resists baffling the less technical reader or sending the technical reader off down another path that will distract but not necessarily inform.

The need to find answers to questions that only simulation can begin to answer will continue to grow exponentially as, other than constructing and operating a building, only simulation can predict how a building might perform in reality. This book will be indispensable not only to students and researchers but also to practitioners, allowing them to gain the skills required to answer complex questions alongside an appreciation of the importance of good choices on outcomes.

Professor Lori B McElroy MBE, PhD, MA, MCIBSE, HonFRIAS, FIBPSA
Department of Architecture, University of Strathclyde
& President IBPSA
August 2020

Preface

> The teaching of building performance simulation (BPS) is a topic that deserves as much attention as the development and validation of models and simulation tools.
>
> As stated by Clarke (2001): *"What is the point of developing powerful tools without putting in place the means to train and support users?"*

Objectives of this book

The International Building Performance Simulation Association (IBPSA) published a *position paper* in 2015 that provides a vision for BPS. This identified the need for a core teaching and learning package that is tool generic (Proposition 15 of Clarke, 2015). The objective of this book is to make a contribution towards this core package.

This book is aimed at teaching the fundamentals of BPS. Readers who complete all the learning elements (described below) will be able to :

1. Understand the models that have been implemented into BPS tools for treating the significant heat and mass transfer processes.
2. Appreciate the simplifications inherent in these models and the necessity for these simplifications.
3. Comprehend the implications of modelling choices and default modelling methods and input data.

4. Select appropriate models, simulation methods, and BPS tools for a given analysis.

5. Understand which modelling choices and input data have the greatest impact upon simulation predictions.

An experiential teaching approach

An article appearing in the *Journal of Building Performance Simulation* (Beausoleil-Morrison, 2019) describes an approach for teaching the fundamentals of BPS. This book has been written to support this teaching approach and is designed to be used as a textbook for one or two semester-long post-graduate courses in engineering, building physics, or architectural science. It can also be used as a self-study guide by BPS practitioners wishing to deepen their knowledge of the fundamentals. It presumes basic knowledge of heat transfer, thermodynamics, building physics, and the terminology of buildings and heating, ventilation, and air-conditioning (HVAC) systems.

The teaching approach consists of four interrelated modes of learning: *Study theory*, *Simulation exercise*, *Simulation autopsy*, and *Reflect & connect*. The result is a learning spiral wherein the completion of one topic's cycle through the four learning modes leads into the next topic. Most of these topics focus on an individual heat or mass transfer process, such as longwave radiation from external building surfaces, convection heat transfer at internal surfaces, air infiltration, etc.

Through the book's first required reading (more on this below) you will become familiar with the learning spiral, the four interrelated modes of learning, and other details of this teaching approach.

Book organization

Each chapter contains text to introduce a topic or to present the essence of theory. Some chapters are long, while others are quite short. This is followed by a required reading or two. As the name implies, these readings are requisite for understanding the chapter so don't skip over them! Most are journal articles or conference papers, and guidance is provided on which aspects to focus on. The reference list contains links to all of the required readings, many of which can be downloaded for free. Although subscriptions are required to access most of the journal articles, readers who do not already have access can usually request a copy by directly contacting the authors. Most chapters also contain a list of sources recommended for further learning for readers wishing to delve into topics to a greater depth.

The first part of the book briefly introduces BPS, defining what it is, how it is used, and discusses the central role the user plays in ensuring valid BPS predictions. Chapter 1 includes the first simulation exercise, in which you will create a one-zone representation of a simple building. In this manner, you will begin to learn and apply your chosen BPS tool in parallel to studying theory.

Each of the next three parts of the book contains a series of chapters. Each distinct heat or mass transfer process is treated by a dedicated chapter appearing in one of these three sections. Part II treats the heat and mass transfer processes relevant to the building interior, while Part III focuses on heat transfer processes relevant to the exterior environment. Heat and mass transfer occurring through the building envelope are the subjects of Part IV. This is followed by Part V, which focuses on HVAC systems.

The structure of each of the chapters appearing in these parts of the book is similar. Basic theories are first introduced and then the methods commonly used in BPS are described. This is done in a tool-agnostic manner whereby the spectrum of commonly employed techniques is outlined, and the strengths and weaknesses of each are described. Mathematical descriptions are provided where necessary to illustrate concepts, but these chapters are not meant to be a comprehensive compendium of models.

The final part of the book includes a *Culminating Trial* in which you will apply all the knowledge and skills you have developed in the preceding chapters. You will represent an actual building with your chosen BPS tool and compare your simulation predictions to measurements.

Simulation exercises and autopsies

Each chapter includes hands-on simulation exercises. These have been carefully formulated to illustrate the chapter's topic and to help students concretize theoretical concepts. These can be conducted with any BPS tool of your choosing, and have been designed so that you will develop skills at applying your chosen BPS tool and at extracting results.

You might want to consult the Building Energy Software Tools directory to help you choose a BPS tool to use in conjunction with this book. It is best to opt for one that is supported with extensive technical documentation so that you can learn not only how to operate the tool, but also understand and control its calculation algorithms. This is important in terms of reinforcing the theory presented in the book. Consider using two different BPS tools as this will provide greater opportunities to examine and contrast the performance of alternate calculation approaches.

You can expect to refer back to the chapter's text and required readings while performing the simulation exercises, each of which requires you to predict specific aspects of performance. To maximize the learning value it is highly recommended that course instructors lead students through a *simulation autopsy* upon the completion of each chapter's simulation exercises. These sessions are invaluable and serve to reinforce the theory, with each student learning not only from their own experiences, but from those of their peers as well. Beausoleil-Morrison (2019) describes by way of examples the organization and learning outcomes of simulation autopsies.

The result of the first simulation exercise in Chapter 1 will be your *Base Case* that you will use in all subsequent simulation exercises in the book (apart from the *Culminating Trial*), most of which will have you perturbing a single BPS input or altering a single modelling approach.

A note to instructors

If you are interested in adopting this book as a text for a new or existing course, I would be happy to share information such as course outlines, lecture materials, video sequences, and tips on structuring simulation autopsies. Just reach out.

The simulation exercises have been successfully tested using a number of BPS tools. Feel free to prescribe a tool to your class that you are familiar with, or allow your students to choose. I have my students conduct the simulation exercises with two different BPS tools, as there is great learning value in contrasting approaches. Do be aware that not all exercises can be run with all tools, but there is a learning opportunity in this. When this occurs you can explain to students why particular tools do not support certain modelling approaches.

The simulation exercises presume no prior knowledge in BPS or a particular BPS tool so don't worry about choosing a tool unfamiliar to the students. It is easy to train new users to operate a BPS tool. I post video sequences of the tools I use in my teaching to support this initial learning. The real challenge comes in learning how to effectively apply a tool with full knowledge of its applicability, modelling limitations, and default methods and data, and in developing skills to scrutinize results. That is what this book is about.

Finally, I encourage you to read the article that describes my teaching approach (Beausoleil-Morrison, 2019) because this book has been designed around it.

Additional resources

The book's companion website contains additional resources for readers and instructors, many of which are referred to in the text. This includes weather data necessary to conduct the simulation exercises, links to relevant tools and websites, and photographs, drawings, and measured data to support the *Culminating Trial*.

Acknowledgements

I would like to thank the many students at Carleton University's *Building Performance Research Centre* who graciously agreed to review and comment on large sections of the book's manuscript. Their insightful feedback and attention to detail were invaluable in improving the clarity and accuracy of the manuscript's presentation.

Special thanks to Sarah Brown, Lumi Dumitrascu, Calene Treichel, Dan Stalinski, Rebecca Pinto, Brodie Hobson, Jayson Bursill, and Joe Coady.

The many valuable comments that were provided by colleagues, including Liam O'Brien, Burak Gunay, and Michaël Kummert, are greatly appreciated. I am grateful for the many helpful suggestions provided by Malcolm Cook, Ruchi Choudhary, and Erik Kolderup, who examined the book's manuscript on behalf of IBPSA. I am also indebted for the ceaseless support provided by Chloe Layman, my editor at Routledge Taylor & Francis Group.

Finally, I wish to thank my dear wife Rachel for her unending support, encouragement, wisdom, and guidance, without which this book would not exist.

Ian Beausoleil-Morrison
Ottawa, May 2020

Nomenclature

Latin letters

A	Area (m^2)
$\mathcal{A}, \mathcal{B}, \mathcal{C}$	Regression coefficients or functions
$\mathcal{A}(p), \mathcal{B}(p), \mathcal{C}(p)$	Laplace solution functions
$\mathcal{A}_I$	Flow coefficient for component I ($kg/s\ Pa^{\mathcal{B}_I}$)
a, b	Incidence angle correction factors
a_i, b_i	Weighting factors
a_{wind}, b_{wind}	Wind boundary layer parameters ($-$)
$\mathcal{B}_I$	Flow coefficient for component I ($-$)
C	Air leakage coefficient ($m^3/s\ Pa^n$)
C_a	Cloud correction factor ($-$)
C_D	Discharge coefficient ($-$)
C_m	Thermal capacitance (J/K)
C_P	Pressure coefficient ($-$)
c_P	Specific heat ($J/kg\ K$)
DHI	Diffuse horizontal irradiance (W/m^2)
DNI	Direct normal irradiance (W/m^2)
E	Emissive power (W/m^2)
$\mathcal{F}$	Shading factor ($-$)
$f(p)$	Subsidiary equation in the Laplace domain
$F(t)$	Function in time domain
$f_{motor \rightarrow a}$	Fraction of heat generation by motor that is added to air ($-$)
f_{stack}	Air leakage stack factor ($-$)

f_{wind}	Air leakage wind factor $(-)$
F_1	Circumsolar brightening coefficient
F_2	Horizon brightening coefficient
$f_{i \to j}$	View factor from surface i to surface j $(-)$
G	Irradiance (W/m^2)
g	Gravitational acceleration (m^2/s)
G_λ	Spectral irradiance (W/m^2 μm)
GHI	Global horizontal irradiance (W/m^2)
H	Height above datum point (m)
h	Enthalpy (J/kg)
h_{conv}	Convection coefficient (W/m^2 K)
h_{lw}	Longwave radiation coefficient (W/m^2 K)
HHV	The higher heating value of a fuel (J/kg)
I_λ	Spectral intensity (W/m^2 sr μm)
k	Thermal conductivity (W/m K)
K_θ	Incident angle modifier at incident angle θ $(-)$
L	Length scale (m)
$\mathcal{L}$	Laplace operator
$LMTD$	Log mean temperature difference
m	Mass (kg)
$\dot{m}$	Mass flow rate (kg/s)
N	Amount of sky dome obscured by opaque clouds (tenths)
n	Air leakage coefficient $(-)$
NTU	Number of transfer units $(-)$
Nu	Nusselt number $(-)$
P	Pressure (Pa)
p	Operator variable in Laplace domain
$\dot{P}_{el}$	Rate of electricity consumption (W)
P_L	Surface perimeter (m)

PLR	Part load ratio (−)
Pr	Prandtl number (−)
q	Rate of heat transfer (W)
R	Steady-state thermal resistance ($m^2\,K/W$)
$\mathcal{R}_f$	Roughness factor (−)
Ra	Rayleigh number (−)
Re	Reynolds number (−)
S_w	Air leakage shelter coefficient (−)
T	Temperature (°C or K)
t	Time (s)
U	Overall heat transfer coefficient ($W/m^2\,K$)
u	Internal energy (J/kg)
V	Velocity (m/s)
W	Rate of work (W)
$\mathcal{W}_f$	Windward/leeward factor (−)
$\mathcal{W}_i, \mathcal{X}_i, \mathcal{Y}_i, \mathcal{Z}_i$	Transfer functions
$X_i, Y_i, \text{and } Z_i$	Response functions
x, y, z	Distances (m)
Z	Heat exchanger heat capacity ratio (−)

Greek letters

α	Absorptivity (−)
β	Surface slope (°)
β_v	Volumetric thermal expansion coefficient (1/K)
χ, ψ	Spherical coordinates (rad)
Δt	Duration of simulation timestep (s)
Δx	A distance (m)
δ	Solar declination angle (°)
ϵ	Emissivity, or heat exchanger effectiveness (−)

η	Efficiency ($-$)
η_{alt}	Solar altitude angle ($°$)
γ	Surface azimuth angle ($°$)
λ	Wavelength (μm)
μ	Viscosity ($\mathrm{N\,s/m^2}$)
Ω	Solar hour angle ($°$)
ω	Humidity ratio ($\mathrm{kg_v/kg_a}$), or angular velocity ($\mathrm{rad/s}$)
Φ	Latitude ($°$)
Ψ	Heat flux in Laplace domain
ρ	Reflectivity ($-$), or density ($\mathrm{kg/m^3}$)
σ	Stefan-Boltzmann constant ($\mathrm{W/m^2\,K^4}$)
τ	Transmissivity ($-$)
Θ	Temperature in Laplace domain
θ	Angle of incidence between solar beam and surface normal ($°$)
ζ	Insolation factor ($-$)
Z	Solar diffuse distribution factor ($-$)

Superscripts

$'$	Moist air property
rated	Evaluated at rated conditions
a0	Absorption without inter-surface reflection
a1	Absorption after one set of inter-surface reflections
r	Reflected radiation
t0	Irradiance transmitted without reflection
t1	Irradiance transmitted after one set of inter-reflections

Subscripts

$\downarrow$	Downwelling
$-$	Horizontal
$\perp$	Normal to the solar beam
a	Dry air
abs	Absorbed
atm	Atmospheric
b	Blackbody
$beam$	Beam radiation
$boiler$	Boiler
c	Solar collector
con	Condenser
$cond$	Conduction
$conv$	Convection
$diff$	Diffuse radiation
dp	Dew point
e	External surface
$e1 \Rightarrow 7$	Total emission from surface 1 that is absorbed by surface 7
$e1$	Emission from surface 1
$east, west$	Boundaries of finite difference nodes
env	An object in the exterior environment that exchanges longwave radiation with the building
$evap$	Evaporator
exf	Exfiltration
fan	Fan
fg	Vaporization from fluid to gas
$forced$	Forced convection
$fuel$	Fuel
g	Grey surface

gas	Gas fill
grd	Ground
HP	Heat pump
HVAC	Heating, ventilation, and air-conditioning
HX	Heat exchanger
$i \rightarrow z$	From *i* to *z*
i	Internal surface
in	Entering zone
inc	Incident
inf	Infiltration
latent	Latent
loss	Loss from a component to the surroundings
lw	Longwave radiation
m	Mass of envelope assembly
max	Maximum
min	Minimum
motor	Motor
natural	Natural convection
oa	Outdoor air
obj	Surrounding objects in the exterior environment
open	Opening
out	Exiting zone
P, E, W	Finite difference nodes
RA	Return from zone under consideration to air-based HVAC system
refl	Reflected
s	A surface other than the one under consideration
SA	Supply to zone under consideration from air-based HVAC system

sky	Atmospheric gases, aerosols, and clouds participating in longwave radiation exchange, as well as deep space
sky_dome	The hemisphere formed by the sky dome
solar	Solar radiation
source	Source within zone
stack	Stack effect
strata	Strata
t − in	Transfer air flowing into zone under consideration from another zone
t − out	Transfer air exiting zone under consideration and flowing to another zone
tank	Tank
trans	Transmitted
v	Moisture (water vapour)
w	Water
wind	Wind
z	Zone

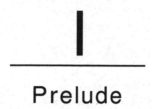

Prelude

Introduction to BPS

THIS chapter provides an introduction to BPS. It briefly describes what it is, how it works, and who uses it. You will come to understand how to use this book, and you will apply your chosen BPS tool to represent a simple building called the *Base Case* that will form the basis for simulation exercises you conduct in subsequent chapters.

Chapter learning objectives

1. Understand the key characteristics of BPS and the role it can play in the design, analysis, and operation of buildings.
2. Appreciate how history has influenced the choice of models utilized in current BPS tools.
3. Become aware of the methods used to validate BPS models and tools.
4. Understand the learning approaches employed in this book.
5. Become familiar with operating your chosen BPS tool and develop skills at translating the description of a simple building into appropriate input data.

1.1 WHAT IS BPS?

BPS employs a large number of mathematical models to simulate a building's performance under a given set of boundary conditions. Many aspects of performance might be appraised by BPS, including energy consumption, ventilation effectiveness, thermal comfort, lighting quality, etc. The objective is to represent the significant physical processes so that the simulation provides an accurate— or at least a useful—representation of reality.

The mathematical models employed in BPS are simplified descriptions of complex systems and processes. They necessarily make approximations to reduce the complexity to a manageable level for both the computer and the user. The necessity of these approximations can be appreciated by focusing on the building illustrated in Figure 1.1.

Figure 1.1: The Urbandale Centre for Home Energy Research (a research facility at Carleton University)

Consider all of the heat and mass transfer processes occurring between this building and its surrounding environment. Heat is being transferred from the warm indoor surfaces of the building envelope to the cold exterior surfaces by means of conduction through solid materials such as the gypsum interior wall board, the wood studs forming the wall's structure, and the cedar cladding. And heat is being transferred by convection, radiation, and conduction (solid and gaseous) through the wall's fibreglass batt and foam insulations. All of these processes are three-dimensional and transient in time.

There is convective heat transfer from the exterior wall surfaces to the outdoor air, which is dependent upon wind velocity (speed and direction). The wind velocity patterns in the vicinity of the building are in turn affected by the house's shape and size, and by surrounding buildings and objects. There is radiation heat transfer in the infrared spectrum from the exterior wall surfaces to water molecules in the earth's atmosphere, to deep space, and to the surfaces of the surrounding ground and objects.

Solar radiation—some of which is scattered by the earth's atmosphere—is incident upon exterior wall and window surfaces. Some of this radiation will be transmitted through window glazing layers, while some will be absorbed and reflected by the individual glazing layers, all of which depends upon the solar radiation's angle of incidence. And some solar radiation is reflected by the ground towards the building, further increasing solar transmission to the interior, but the amount of reflection is dependent upon the composition of the snow cover, which is influenced by moisture content, temperature, and time.

The solar collectors on the roof of the building are partially covered from a fresh snowfall, which influences their ability to capture solar gains, and therefore impacts the thermal storage and auxiliary heating systems. The positioning of the blinds is another complication, as they affect the previously mentioned infrared and solar radiation processes, and may be controlled by occupant behaviour. Air can infiltrate past the house's air barrier due to imperfections in sealing, and these flows will depend upon local indoor–outdoor pressure differences in the vicinity of these unintentional openings.

This is only a partial inventory of the significant heat and mass transfer processes occurring in reality. Mathematical models must be constructed to represent each significant physical process and these must be discretized in numerical form, and then solved collectively. This solution approach is necessary because all of these processes are interconnected so they cannot be solved in isolation. For example, the solar radiation absorbed on the exterior glazing surface influences its temperature which, in turn, influences the infrared radiation emitted by that glazing to the atmosphere.

It is easy to understand why simplifications are necessary to reduce this complexity to a manageable level of detail for solution purposes. But this necessity to simplify is also driven by user considerations. Can you imagine how a user might describe when blinds would be retracted or deployed? Or whether snow will build-up on the solar collectors, and how quickly it might melt?

Due to the complexity of the reality—and the necessity of simplifications—BPS inherently operates with significant uncertainty, and this must be recognized and acknowledged by users. How accurately can the solar reflectivity of that snow be estimated, and what impact does this have upon simulation predictions? What is the thermal conductivity of the wood studs within the wall assemblies (did the construction crew use fir or spruce?), and what impact does this uncertainty have upon predictions?

Another characteristic of BPS is that it operates with transient (time-varying) boundary conditions. We rarely wish to predict the performance of buildings at a single snapshot in time. Rather, we typically use BPS to march through time (maybe a week, a month, a year, or multiple years) to predict performance subject to time-varying boundary conditions. These time-varying boundary conditions include occupant presence and behaviour (e.g. window openings, appliance operation) and changing weather, all of which must be prescribed by the user or predicted with models.

1.2 HOW DOES IT WORK?

The user must provide considerable input data to the BPS tool to exercise all of these models. The level of detail required depends upon the modelling approaches employed by a particular tool or selected by the user. Table 1.1 broadly categorizes the types of input data that may be required.

Table 1.1: Types of input data required from user

Category	Inputs
Geometry	Building plan and elevation
	Internal space layout
	Window sizes, locations, and shades
	Shading by neighbouring buildings and objects
Materials	Properties of structural and insulating materials
	Radiative properties of glazings
HVAC	Energy conversion and distribution systems
	Ventilation systems
	Component and supervisory controls
Airflow	Window and other intentional openings
	Cracks, holes, and defects in air barrier
	Airflow paths between internal spaces
Internal gains	Electrical appliances and lighting
	Moisture sources, such as cooking and plants
Occupants	Occupant density and schedule
	Activities that generate heat and moisture
	Control of appliances and lighting
	Interactions with windows and thermostats
Weather	Solar radiation
	Air temperature and humidity
	Wind speed and direction
	Sky conditions
	Ground snow cover
	Microclimate effects

The user must make choices about how much resolution to include, and this will dictate the amount of detail required in the form of inputs. These decisions should be driven by the objectives of the simulation study. For example, if the goal is to predict energy consumption or peak indoor air temperatures, then it would be appropriate to represent the *significant* heat and mass transfer processes that affect the building's thermal performance. It may not be necessary to represent some features of the building to accomplish this. In many cases such a goal can be achieved with a highly abstracted representation that neglects many of the building's geometrical details.

Some thermal and mass transfer processes are critical in some situations, but have minimal influence in others. Geometrical details matter in some cases, but have a negligible influence in others. There are no simple rules. The user must choose which models and how much detail to include. As each BPS tool embodies only a subset of available models, the selected tool will dictate many of these choices. For this reason, careful planning is of paramount importance.

1.3 WHAT IS IT USED FOR?

BPS is used by engineers, architects, building physicists, and researchers in many domains of application. Predictions (outputs) can be provided in varying levels of detail in a number of different categories, although it must be stressed that not all tools provide predictions for all domains.

BPS can be deployed at many stages of a building's life cycle. It is often used to research novel energy conversion and storage systems for buildings. It can be used to help establish the shape, size, and layout of a building (pre-design and schematic design phases). And it can help detail the design of the building envelope, HVAC, and lighting systems (design development phase). BPS is often used post-design to demonstrate compliance with building or energy regulations, or as a requirement of energy labelling programmes. Although less common, it can also be used to assist during building commissioning and to improve building operations (building controls, fault detection).

A partial listing of the applications of BPS is provided in Table 1.2.

1.4 BPS TOOLS AND USERS

Many BPS tools have been developed over the past half century[i]. Some are targeted at researchers, while others at practitioners. Some tools attempt to consider all physical processes relevant to buildings, while others focus on thermal, lighting, indoor air quality, or other areas. Some are complex and present users with myriad choices of modelling options, while others present simplified user interfaces and apply many underlying assumptions and implement default methods and input data.

Each tool has strengths and weaknesses. To become a BPS expert, you must understand the underlying methodologies employed by tools and develop the expertise to employ multiple tools, as no single tool meets the needs of every situation.

Most BPS practitioners are engineers and architects. The BPS field is emerging as a specialization in the building industry, but it must be pointed out that there is no licensed BPS profession as such. The accreditation of BPS users is a topic that is receiving greater attention in many jurisdictions.

The BPS field has been rapidly changing in the past decade, but unfortunately the application of BPS is integral in the design of only a small minority of buildings. Most applications are for post-design evaluation, to demonstrate compliance with regulations, or to qualify for voluntary labelling programmes.

1.5 A BRIEF HISTORY OF BPS

Understanding some early BPS history is important, because decisions taken decades ago have determined some of the approaches still in use today.

---

[i]The Building Energy Software Tools directory provides a comprehensive listing of available tools.

Table 1.2: Applications of BPS

Category	Prediction
Thermal	Predicting energy consumption
	Estimating peak heating and cooling loads
	Sizing HVAC equipment
	Assessing building form and fabric
	Examining external shading
	Determining overheating risks
	Comparing HVAC systems
	Assessing natural and hybrid ventilation
	Exploring novel energy systems
Indoor environment	Ventilation effectiveness
	Airflow distribution
	Indoor air quality
	Daylighting
	Lighting quality
	Thermal comfort
Operations	Fault detection
	Model predictive control
	Comparing control options
Other	Occupant behaviour and movement
	Coupled heat, air, and moisture transfer
	Acoustics
	Fire propagation
	Building evacuation
	External airflow

Early approaches

Until the mid 1960s only simple hand-calculation methods were available for estimating energy usage in buildings. These included the *degree day* method and the more detailed *bin* method, both of which are still used.

Degree days, a measure of a climate's severity, are calculated by integrating over the year the daily-averaged outdoor-air temperature relative to a fixed base (most commonly 18 °C). Degree days for various locations were tabulated, published, and used in conjunction with the steady-state peak heating load and a fixed heating-system efficiency to estimate the usage of heating fuel over the year. Although easy to apply, the degree day method neglects many significant factors, such as transient thermal storage in building materials, solar gains, internal gains, variations in ventilation and infiltration rates, and the non-steady operation of heating equipment.

As with the degree day approach, the bin method treats outdoor air temperature as the independent variable in the analysis. The analysis period—usually a year in that era—is sorted into bins according to the outdoor temperature. Each bin thus contains the number of occurrences (usually measured in hours) within its range of outdoor temperatures (typically ∼3 °C wide). The energy consumption of each bin is determined (independently) using simplified steady-state approaches much like those of the degree day method. The predictions from all bins are then summed, yielding an estimate of the building's heating and cooling energy consumption.

Compared to the degree day approach, the bin method allows some assumptions about fixed conditions to be dropped: infiltration rates and cooling system efficiencies can vary with indoor–outdoor temperature difference, for example. However, the bin method implicitly assumes that energy flows within the building are exclusively a function of indoor–outdoor temperature difference. Therefore the timing (even day versus night) of solar and internal gains, and transient indoor conditions cannot be explicitly considered. Although more resolved binning approaches have been introduced in an attempt to address this fundamental shortcoming, the unifying characteristic of all bin methods is that time has been eliminated as a variable in the analysis.

True simulation methods

The first true simulation methods—true in that they attempted to imitate physical conditions by treating time as the independent variable—appeared in the mid 1960s (GATC, 1967).

Because computing resources were limited, slow, and extremely expensive, it was necessary to subdivide the problem domain. The so-called *Loads-Systems-Plant* (LSP) modelling strategy was commonly employed in these early approaches. It subdivided the simulation of the building into three sequential steps. The building's heating and cooling loads are first calculated for the entire analysis period (often a year) for an assumed set of indoor environmental conditions. These loads are then imposed as inputs to the second step of the simulation, which models the HVAC system's air handling and energy distribution components (fans, heating coils, cooling coils, air diffusers, etc.). This second simulation step (also conducted for the entire analysis period) predicts the demands placed on the HVAC system's energy conversion components (boilers, chillers) and related equipment (cooling towers and circulation pumps). Finally, the energy conversion and related systems are simulated in the third step, receiving as input the results of the second step.

Obviously, the sequential nature of the LSP approach neglects interactions between the steps. The impact of undersized heating or cooling equipment cannot be considered. Furthermore, situations in which there is strong coupling between the steps (e.g. the impact of the air handling system on infiltration; the impact of room temperatures on occupant behaviour such as the opening and closing of windows) cannot be adequately treated.

Response factor methods

Many of the early simulation methods utilized simplified approaches for modelling building loads, such as the time-averaging approach, which smeared internal heat gains over a period of time to roughly approximate the transient thermal storage, radiation, and convection processes that were actually occurring.

New techniques were introduced to address such shortcomings.

The pioneering work of Stephenson and Mitalas (e.g. Stephenson and Mitalas, 1967) on the *response factor* method significantly advanced the modelling of transient heat transfer through the opaque fabric and the heat transfer between internal surfaces and the room air. They utilized the principle of superpositioning to decompose the complex non-linear heat transfer system into a summation of responses of the component parts. This allows, for example, solar insolation to be modelled with a simple algebraic summation, using *weighting factors* which relate the convection (of heat to the room air) to the solar radiation absorbed by internal surfaces at previous periods of time. Heat transmission through the walls is calculated by another (independent) summation, this one operating on the time-series history of wall surface temperatures. In effect, the response factor method decouples the treatment of solar insolation from the modelling of heat transfer through walls.

Heat balance approaches

Heat balance approaches were introduced in the 1970s (e.g Kusuda, 1976) to enable a more rigorous treatment of building loads. Rather than utilizing weighting factors to characterize the thermal response of the room air to solar insolation, internal gains, and heat transfer through the fabric, this methodology solves heat balances for the room air and at the surfaces of fabric components.

These heat balances consider all important energy flow paths: transmission through the fabric, radiation exchange between internal surfaces, solar insolation, convection from the indoor air to wall and window surfaces, etc. The heat balances are formed and solved each timestep to estimate surface and room air temperatures, and heat flows. Although more computationally demanding than room-air weighting factors, the introduction of the heat balance approach allowed some significant assumptions of linearity to be dropped. For example, convection coefficients characterizing heat transfer from internal surfaces to the room air could respond to thermal states within the room, rather than being treated as constant.

HVAC and airflow modelling

More complex and rigorous methods for modelling HVAC systems were introduced in the 1980s. Transient models and more fundamental approaches were developed as alternatives to the traditional approach which performed mass and energy balances on pre-configured templates of common HVAC systems, the components of which (fans, coils, boilers, etc.) were represented by overall efficiency values, calculated by curve fits to manufacturers' data. Additionally, in the 1980s the simulation of building loads and HVAC were integrated in order to consider the important interactions between the two.

Activity in the building simulation field was not limited to thermal considerations. Parallel work was underway on airflow modelling. Methods were developed for estimating wind- and buoyancy-driven infiltration rates, and computational fluid dynamics approaches were being applied to simulate the details of airflow patterns within single rooms. In the 1970s, network airflow models were developed for simulating both infiltration and internal airflow. These are macroscopic models, which represent large air volumes (e.g. rooms) by single nodes, and predict flow through discrete paths (e.g. doors, cracks).

The thermal and airflow simulation approaches did not begin their convergence until the mid 1980s (e.g. Walton, 1983), at which time the network airflow models were integrated into thermal models to couple the simulation of heat and airflow, and to analyze pollutant dispersion within buildings. Until this time, the thermal simulation tools focused strictly on energy processes. Although the thermal impact of both air infiltration and (in some cases) inter-zone airflow was considered, flow rates were either user-prescribed or estimated using simplified approaches. Airflow was not simulated, but rather merely its impact considered in the thermal simulation. As a result, configurations in which heat and airflow were strongly coupled (e.g. naturally ventilated buildings) could not be accurately simulated.

Parallel developments occurred on simulating many other aspects of building performance, including acoustics, daylighting and lighting, indoor environment quality, etc.

This history continues

Decisions taken during the early decades of BPS's development are still having an impact today. Although rapidly expanding computing power over the past quarter century has eliminated the prime motivating factor for many of the earlier simplifications, all of the aforementioned methods are still in use.

There are BPS tools that are based upon superpositioning and response factors while others employ the heat balance approach. Some model transient conduction with conduction transfer functions while others use numerical methods. Some tools couple thermal and airflow simulations, others treat these domains disparately. Although most BPS tools tightly couple the simulation of building loads and HVAC systems, some are still based upon the segregation of these modelling domains.

Most BPS tools in use today have evolved over long periods of time and are written using procedural computer languages that execute models in a sequence defined by the tool developers. Most of these march through time using fixed simulation timesteps (sometimes under the user's control, sometimes not). More recently, some BPS tools have been developed with equation-based acausal modelling languages to provide greater flexibility in solution approaches (see, for example, Wetter, 2009).

It is not the case that some methods and tools are superior to others. Each has its place and has a certain applicability. But it is important for the user to understand the underlying methods employed by each tool so that informed choices can be made about which tool to employ in which situation, and to choose appropriate methods within a given tool.

1.6 IS BPS VALID?

The validation of BPS tools and models is a complex and challenging field that has existed almost as long as BPS itself. Extensive efforts have been conducted under the auspices of the International Energy Agency (IEA), the American Society for Heating Refrigeration and Air-Conditioning Engineers (ASHRAE), the International Organization for Standardization (ISO), and others to create methodologies, tests, and standards to verify the accuracy and reliability of BPS.

These validation initiatives have proven effective at diagnosing some so-called *internal sources of errors* in BPS models and tools. According to Judkoff *et al.* (1983) these internal sources of errors can be classified as follows:

- Differences between the actual thermal transfer mechanisms taking place in the reality and the simplified model of those physical processes in the simulation.
- Errors or inaccuracies in the mathematical solution of the models.
- Coding errors.

Judkoff and Neymark (1995) proposed a pragmatic approach composed of three primary validation constructs to check for these internal errors. These are:

- Analytical verification.
- Empirical validation.
- Comparative testing.

With analytical verification, BPS outputs are compared to a well known analytical solution for a problem that isolates a single heat transfer mechanism. Although analytical verification is limited to simple cases for which analytic solutions are known, it provides an exact standard for comparison.

BPS outputs are compared to monitored data with empirical validation. The measurements can be made in real buildings, controlled test cells, or in a laboratory. The design and operation of experiments leading to high-quality data sets is complex and expensive, thus restricting this approach to a limited number of cases. The characterization of some of the more complex physical processes (such as heat transfer with the ground, air infiltration, indoor air motion, and convection) is often excluded due to measurement difficulties and uncertainty.

A BPS tool is compared to itself or other tools with comparative testing. This includes both sensitivity testing and inter-program comparisons. This approach enables inexpensive comparisons at many levels of complexity. However, in practice the difficulties in equivalencing program inputs can lead to significant uncertainty in performing inter-program comparisons.

A general principle applies to all three validation constructs. The simpler and more controlled the test case, the easier it is to identify and diagnose sources of error. Realistic cases are suitable for testing the interactions between algorithms, but are less useful for identifying and diagnosing errors. Although the comparison of the actual long-term energy usage of a building with simulation results is perhaps the most convincing evidence of validity from the building designer's perspective, this is actually the least conclusive approach. This is because the simultaneous operation of all possible error sources combined with the possibility of offsetting errors means that good or bad agreement cannot be attributed to program validity (e.g. Lomas *et al.*, 1997).

These painstaking validation efforts are important and do lead to improved BPS tools and greater confidence in the underlying models. That said, it must be recognized that the user is, by far, the largest source of error—not the tools or their underlying models. (You will discover some evidence of this during this chapter's required reading.) Data entry errors can and will occur, and these can go undetected, especially if results are insufficiently scrutinized. But more importantly, the inappropriate use of default models and data, or the selection of inappropriate modelling options for the case at hand are contributing factors. This is why an understanding of the fundamentals is so important.

1.7 REQUIRED READING

Reading 1–A

Beausoleil-Morrison (2019) describes an approach for teaching BPS through a continuous learning spiral that includes exposure to theories and the application of tools. Read this article in its entirety and find answers to the following questions:

1. What are some of the common causes of inaccurate BPS performance predictions (for both experienced and novice users)?

2. What are the learning spiral's four interrelated modes of learning?
3. What is a simulation autopsy?
4. What were some of the causes of the outlying results predicted for the culminating trial reported in Section 7?

1.8 SOURCES FOR FURTHER LEARNING

- The preface and first chapter of Hensen and Lamberts (2019) introduces BPS, describes how it is currently employed throughout the building life cycle, and describes some of the barriers impeding greater adoption of the technology.
- Chapter 1 of Clarke (2001) provides an overview of the objectives of BPS, describes the major energy and mass transfer processes to be modelled, and introduces some of the principal modelling methods that have been devised for resolving them.
- Clarke and Hensen (2015) discuss the role BPS can play in the design and operation of energy-efficient buildings and review the progress the domain has made towards achieving the goal of providing practitioners the ability to accurately and rapidly appraise building design variants for a range of performance metrics. They also treat the shortcomings of the state-of-the-art and list a number of development priorities for the domain.
- Crawley *et al.* (2008) survey a number of BPS tools and contrast their capabilities.
- Kusuda (1999), Sowell and Hittle (1995), and Ayres and Stamper (1995) provide an historical overview of the BPS field and describe the significance of some of its early developments, although principally from an American perspective. It is interesting to contrast the observations made by these authors in the 1990s to the more contemporary perspectives of Clarke and Hensen (2015).
- Clarke (2015) summarizes the requirements of so-called *high-integrity* BPS and details the future developments required to realize this objective.

1.9 SIMULATION EXERCISES

As the topic of this chapter is an introduction to BPS, this chapter's exercises are meant as an introduction to your chosen BPS tool. They will be, by far, the most time-consuming simulation exercises to conduct apart from those of Chapter 18. Expect some frustrations and minor failures: these are an important part of the learning process!

Objectives

The objectives of this first exercise are for you to become familiar with your chosen BPS tool and to develop skills at translating a building description into appropriate input data. The subject is a simple building with a rectangular floor plan and a single window which includes many simplifications so that you can focus on learning the basics. Normally the user of a BPS tool must make many modelling decisions and assume some input data, but in this case a complete description of the building is provided.

The BPS representation you create here will form the *Base Case* for the simulation exercises that will be performed in subsequent chapters, each of which will explore the impact of modelling choices and input data related to a single heat or mass transfer process or aspect of an HVAC system. Therefore, it is important that you develop as accurate a representation of the *Base Case* as possible with your BPS tool in order to maximize the learning value of these subsequent simulation exercises.

Geometry

The geometry of the building is illustrated in Figure 1.2. The building measures 12.2 m by 6.1 m by 2.7 m high, with its longer dimension aligned east–west. There is a single window measuring 3 m wide by 1.75 m high. It faces south and is mounted at the location indicated in the figure.

Internal partition walls and furnishings can be ignored. Construct a one-zone representation of this building in your chosen BPS tool.

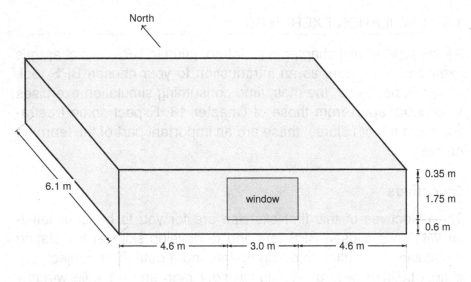

Figure 1.2: *Base Case* geometry

Opaque envelope constructions

The construction of the walls, floor, and roof are provided in Tables 1.3 through 1.5. Assume each material layer to be homogeneous, and ignore the thermal impact of building structural components, fasteners, and other thermal bridges.

The solar absorptivity of the gypsum board used in the wall and roof assemblies is 0.6. The floor's concrete has a solar absorptivity of 0.6, while the wall's cedar and the roof's asphalt have values of 0.78 and 0.85, respectively. The emissivity of all construction materials is 0.9 in the longwave portion of the infrared spectrum.

Window

Data are provided in this section to support all of the models that are currently used by BPS tools for calculating the radiation, convection, and conduction processes that occur across windows. From this, you will have to select the input data required by your chosen BPS tool and model.

The window, which remains closed at all times, is triple-glazed and filled with argon. The outer glass layer is a 3 mm thick sheet of clear glass. The inner two layers of glass are also 3 mm thick but

Table 1.3: Wall construction (inside to outside)

Element	Thickness (mm)	Conductivity (W/m K)	Density (kg/m³)	Specific heat (J/kg K)	R-value[a] (m² K/W)
Gypsum board	13	0.16	640	1 150	
Glass fibre	140	0.04	12	800	
EPS[b]	30	0.033	29	1 500	
OSB[c]	13	0.1	800	1 880	
Air gap	50				0.18
Cedar	20	0.1	360	1 630	

[a] Nominal steady-state thermal resistance, or R-value; required input in some BPS tools.
[b] Expanded polystyrene
[c] Oriented strand board

Table 1.4: Floor construction (inside to outside)

Element	Thickness (mm)	Conductivity (W/mK)	Density (kg/m³)	Specific heat (J/kgK)
Cast concrete	100	1.90	2 300	840
XPS[a]	75	0.029	35	1 500

[a] Extruded polystyrene

Table 1.5: Roof (ceiling) construction (inside to outside)

Element	Thickness (mm)	Conductivity (W/mK)	Density (kg/m³)	Specific heat (J/kgK)	R-value[a] (m²K/W)
Gypsum board	13	0.16	640	1 150	
Glass fibre	350	0.04	12	800	
Air gap	100				0.18
Asphalt shingles	5	1.0	920	1 260	

[a] Nominal steady-state thermal resistance, or R-value; required input in some BPS tools.

contain low-emissivity coatings on surfaces 3 and 5, as illustrated in Figure 1.3. The thickness of the argon gas layers is 13 mm. The presence of window frames and glazing spacers is to be ignored. As well, there are no curtains or blinds.

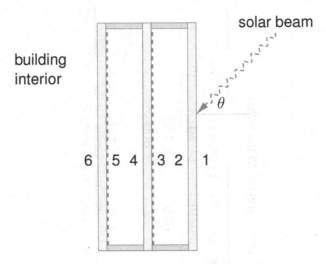

Figure 1.3: Cross-section of *Base Case* window (low-emissivity coatings [indicated by - - - - -] are located on glazing surfaces 3 and 5)

The thermophysical and radiation properties of each glass layer are provided in Table 1.6. The solar radiation properties provided in this table are for when the solar irradiance is normal to the surface of the glass (when the value of θ that is illustrated in Figure 1.3 is zero). Some BPS tools require these data as inputs and then apply models to predict behaviour for off-normal solar irradiance. You will learn about these methods in Chapter 14. Other BPS tools require angularly dependent radiative properties, which are provided in Table 1.7.

Some BPS tools do not support the modelling of the individual conduction, convection, and radiation modes of heat transfer occurring within glazing assemblies, and instead rely on compound performance metrics that express the heat transfer for one set of nominal operating conditions. These metrics are the *solar heat gain*

Table 1.6: Thermophysical and radiative properties of glass layers

Property	Glass layer 1-2	Glass layer 3-4	Glass layer 5-6
Conductivity (W/m K)	1.0	1.0	1.0
Thickness (mm)	3	3	3
Density (kg/m³)	2 500	2 500	2 500
Specific heat (J/kg K)	750	750	750
$\tau_{solar,\perp}$ [a] (-)	0.876	0.708	0.708
$\rho_{solar,front,\perp}$ [b] (-)	0.078	0.179	0.179
$\rho_{solar,back,\perp}$ [c] (-)	0.078	0.155	0.155
τ_{lw} [d] (-)	0	0	0
$\epsilon_{lw,front}$ [e] (-)	0.84	0.095	0.095
$\epsilon_{lw,back}$ [f] (-)	0.84	0.84	0.84

[a] Solar transmissivity at normal incidence.
[b] Front side (side facing sun) solar reflectivity at normal incidence.
[c] Back side (side facing interior) solar reflectivity at normal incidence.
[d] Transmissivity to longwave portion of infrared spectrum.
[e] Front side (side facing sun) hemispherical emissivity in longwave portion of infrared spectrum.
[f] Back side (side facing interior) hemispherical emissivity in longwave portion of infrared spectrum.

Table 1.7: Angularly dependent radiative properties of window assembly and glass layers

Property	Incidence angle (°, θ in Figure 1.3)										
	0	10	20	30	40	50	60	70	80	90	hemisphere[e]
$\tau_{solar,window}$ [a] (-)	0.461	0.461	0.455	0.449	0.441	0.418	0.360	0.249	0.106	0.000	0.383
$\rho_{solar,front,window}$ [b] (-)	0.290	0.290	0.290	0.292	0.300	0.323	0.380	0.503	0.702	1.000	0.357
$\rho_{solar,back,window}$ [c] (-)	0.255	0.255	0.255	0.258	0.267	0.289	0.344	0.463	0.672	1.000	0.322
$\alpha_{solar,1-2}$ [d] (-)	0.057	0.057	0.059	0.060	0.063	0.067	0.072	0.078	0.082	0.000	0.066
$\alpha_{solar,3-4}$ [d] (-)	0.118	0.118	0.121	0.122	0.121	0.120	0.120	0.115	0.080	0.000	0.116
$\alpha_{solar,5-6}$ [d] (-)	0.074	0.074	0.075	0.076	0.075	0.072	0.068	0.056	0.030	0.000	0.068

[a] Direct solar transmission through entire window assembly.
[b] Reflectivity of entire window assembly to solar irradiance incident on front side (side facing sun).
[c] Reflectivity of entire window assembly to solar irradiance incident on back side (side facing interior).
[d] Solar absorptivity of glass layer i-j (refer to Figure 1.3).
[e] Integrated over hemisphere.

coefficient (SHGC[ii]) and the steady-state thermal transmittance coefficient (U-value). The modelling methods that rely on these metrics will be described in Chapter 14. If your chosen BPS tool does not accept the inputs provided in Table 1.6 or Table 1.7, then you can use the glazing assembly's nominal SHGC of 0.585 and U-value of 0.776 W/m² K.

Indoor environment and HVAC

There is a constant sensible internal heat gain of 200 W. These gains are 100% convective. There are no moisture sources within the building (i.e. no latent heat gains).

The building is conditioned by an idealized HVAC system with sufficient heating and cooling capacity to maintain the indoor air temperature between 20 °C and 25 °C at all times. The heat injection and extraction from this idealized HVAC system is 100% convective and 100% sensible.

Exterior environment

The building is located in Ottawa, Ontario, Canada. Use the Ottawa *Canadian Weather for Energy Calculations* weather file in your analysis[iii].

The underside of the floor is exposed to the ambient air. The building is located on a horizontal site in the middle of a field. There are no surrounding obstructions that shade the building.

There is a constant rate of air infiltration to the building of 0.1 ac/h[iv].

[ii]Sometimes referred to as the g-value.

[iii] Use the 2016 version of the CWEC file, not the original 2005 version that is shipped with some BPS tools. The original CWEC weather files were based on weather data from the 1950s to 1990s, whereas the 2016 version is based upon more recent data. The 2016 version can be downloaded from the book's companion website.

[iv]*Air changes per hour* is a commonly used metric for expressing air infiltration and ventilation rates. It is the equal to the volumetric flow rate per hour divided by the building's (or room's) volume.

Exercise 1–A

Create a representation of this building in your chosen BPS tool. Carefully consider and follow all the provided specifications and double-check all your inputs.

Once you are able to complete a simulation, you should scrutinize your results in detail. Examine, for example, zone air temperatures, the rate of solar radiation transmitted through the window, internal heat gains, air infiltration rates, and the rate of HVAC heat injection/extraction. Are these results consistent with your expectations? Don't assume that an absence of error or warning messages from your BPS tool means that your representation is valid or accurate.

Once you are satisfied with your representation of the *Base Case*, perform an annual simulation and extract the following results:

1. Plot the predicted zone air temperature as a function of time. Do the results agree with your expectations?
2. Determine the annual space heating load (GJ).
3. Determine the annual space cooling load (GJ).
4. Determine the magnitude (kW) and timing (day and time) of the peak heating load.
5. Determine the magnitude (kW) and timing (day and time) of the peak cooling load.

Exercise 1–B

If your chosen BPS tool allows you to control its simulation timestep, then perform a series of simulations to explore the impact of this user choice. Examine various timesteps ranging between 1 minute and 1 hour.

What impact does the timestep have upon the predicted annual space heating load? And upon the annual space cooling load? Hypothesize an explanation for your findings.

If your chosen BPS tool does not allow you to control its simulation timestep, then consult its help file or technical documentation to determine why.

1.10 CLOSING REMARKS

This chapter defined BPS, explained how it is used, and discussed the central role the user plays in ensuring valid BPS predictions. Some historical background on the development of BPS models and tools was also provided. The learning approach that is supported by this book was made clear through the chapter's required reading.

This chapter described some of the significant heat and mass transfer processes that have to be considered. It explained that models are required to resolve each one of these in order to predict the performance of a building using BPS. Through the simulation exercises, you caused your chosen BPS tool to make use of many such models in order to predict the performance of the *Base Case*. The next chapter begins to look at how your BPS tool used the information you provided.

Building interior

CHAPTER 2

Energy and mass transfers within buildings

T HIS chapter explains how energy and mass balances are formed for thermal zones and internal building surfaces. It is important to understand how these balance equations are configured as the following chapters will treat how each of the terms appearing in these equations are modelled.

Chapter learning objectives

1. Become familiar with the concept of the *thermal zone*, and begin to appreciate the implications of decisions regarding geometrical abstraction and zoning.
2. Understand how mass and energy balance principles are applied within BPS.
3. Become cognizant of all the terms appearing in the energy balances representing zone volumes and internal surfaces.
4. Become aware of the relative magnitude of the predicted heat transfers for a simple situation.

2.1 SIGNIFICANT PROCESSES

Some of the individual heat and mass transfer processes occurring between a building and its surrounding environment were discussed in the previous chapter (see Section 1.1). Examples of transfers of heat by conduction, convection, and radiation were given, as was an example of a mass transfer through the building envelope.

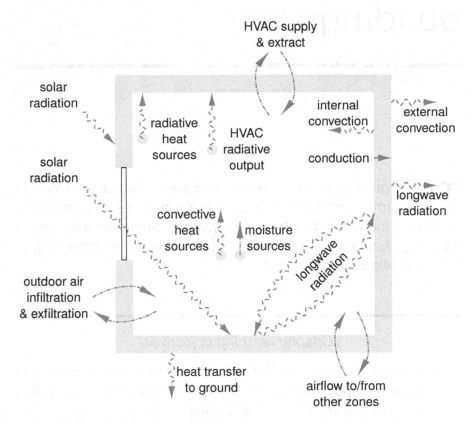

Figure 2.1: Significant heat transfer (〰▶) and mass transfer (⎯⎯▶) processes within buildings

These types of heat transfer and mass transfer processes are illustrated schematically in Figure 2.1 for a very simple building. Each type of process illustrated in this figure can represent multiple individual processes. For example, there will be multiple individual internal convection processes even for the simplest of buildings: from the internal surface of each wall to the surrounding room air, from the upper surface of the floor to the surrounding room air, from the inner surface of the innermost window glazing layer to the surrounding room air, etc.

Some of the processes illustrated in Figure 2.1 occur within the building's interior, some occur between the building's external surfaces and the exterior environment, while others occur through the building envelope. Part II of this book, which includes this and the next few chapters, focuses on the processes occurring within the interior of the building. This chapter describes how energy and mass balances are formed and solved, while Chapters 3 through 6 delve into the details of how each type of process is modelled. In a similar fashion, Part III of the book deals with the exterior environment and Part IV deals with processes occurring through the building envelope.

Before we delve into the formation of energy and mass balances, the important concept of the *thermal zone* will be introduced.

2.2   THE THERMAL ZONE

The building under consideration is subdivided into *thermal zones* for a BPS analysis. This process is known as *zoning* the building. This is one of the first steps in any BPS analysis, and one that should receive considerable attention as choices made at this stage will have a significant impact upon input data requirements, accuracy, and troubleshooting complexity.

A zone represents the air volume and any containing solids of a portion of a building. This may be a single room, a grouping of rooms, the entire building, or a portion of a large space. In the case of a single room, the zone would contain the air volume of the room and any objects within this volume, such as furniture, books, appliances, etc. In the case of multiple rooms or an entire building, the zone would also include internal partition walls and elements of the building's structure (excluding the building envelope). However, as we will see in Section 2.6 the user must usually take some action for the thermal impact of these solid materials to be considered.

The BPS tool will simulate the heat and mass transfers occurring within each zone, and the heat and mass transfers occurring between zones. The air contained within a zone is usually treated as having uniform conditions, so that localized stratification effects are ignored. This is known as the *well-mixed* assumption. Some BPS tools allow more refined treatments, although these are seldom employed. (Chapter 7 will touch upon some of the possibilities.)

It is up to the user to decide how to represent a building's geometry and how to zone the building. There are no hard rules. That said, keep the following general considerations in mind:

- What are the goals of your BPS analysis: predicting energy consumption, studying the impact of building form and fabric, contrasting the performance of HVAC system options, etc.?
- Which heat and mass transfer processes are important for this analysis?
- Do you need to analyze the whole building, or will considering only a portion provide sufficient information?
- Include geometrical features only if they significantly influence heat and mass transfer.
- The zone is a thermal concept, not a geometric one.
- Zones can represent volumes that are non-contiguous.
- Use as many zones as necessary, but as few as possible.

Consider the floor plan of the office given in Figure 2.2. This entire floor of the building could be represented by a single thermal zone. This might be an appropriate zoning strategy for an early-stage analysis if the goal is to predict the annual energy consumption and if the building is conditioned with a single HVAC system. In this case, the entire air volume contained within the building envelope as well as the interior partition walls and all furnishings and other objects would be contained within this one zone.

Another strategy would be to represent each office as its own thermal zone. This treatment could be extended to the reception, meeting room, and corridor. In this case the thermal zones would represent the air volumes of each space and their containing objects, but not the interior partition walls; these partition walls would

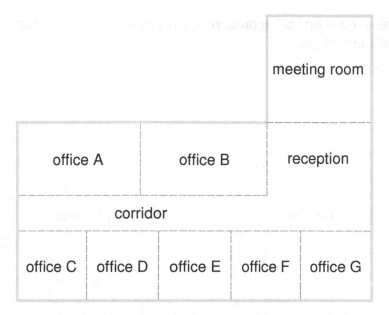

Figure 2.2: The floor plan of an office building (building envelope indicated by ——, interior partition walls by ----)

have to be separately described so that their heat transfer and energy storage processes could be explicitly considered. Obviously more geometrical data would be required from the user to define these 10 zones. Additionally, the user would have to provide sufficient information to characterize airflow from one zone to another (see Figure 2.1) as this would impact the energy balances of the individual zones. Although more input data would be required from the user, the analysis could (potentially) provide a greater resolution of outputs, although it must be recognized that the likelihood of data-input errors would rise.

An intermediate option would be to combine some of the individual spaces into thermal zones. For example, offices A and B might be combined into one zone, offices C, D, E, F, G and the corridor into a second thermal zone, and the meeting room and reception into a third thermal zone. Or, the non-contiguous offices D and F could be combined into one zone, and the non-contiguous offices C, E, and G into another. These four zoning strategies are illustrated in Figure 2.3. (Many other options are, of course, possible.)

There is no "correct" approach: each strategy has its advantages and disadvantages.

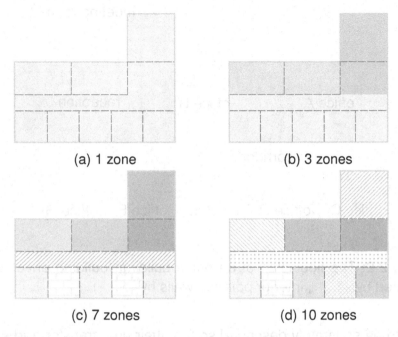

(a) 1 zone   (b) 3 zones

(c) 7 zones   (d) 10 zones

Figure 2.3: Some zoning options for the floor plan of Figure 2.2 (each shade or pattern represents a thermal zone)

The following criteria are recommended when you are considering grouping volumes of the building into thermal zones:

- Volumes should have similar internal loads (magnitude and schedule).
- Volumes should have the same heating and cooling setpoints and thermostat schedules.
- Volumes should have similar solar gains.
- Perimeter volumes should be zoned separately from internal volumes.
- Volumes should not be combined into zones if they are served by different HVAC systems.

Strategies for thermal zoning are best learned through experience. This chapter includes a simulation exercise that will provide you with a first opportunity for exploring the impact of different zoning strategies. A bigger challenge will come in Chapter 18 when you will have to select a zoning strategy in developing a BPS representation of an actual building.

2.3 FORMING ENERGY AND MASS BALANCES

In order to simulate the heat and mass transfer processes occurring within the interior of the building, the BPS tool establishes mass and energy balances for each zone and energy balances for each internal surface within each zone. As will be seen in the upcoming sections, these balance equations contain many terms and are mathematically coupled. Consequently, the user's decisions on zoning strategies and geometrical abstraction have a significant impact upon the number of equations that will have to be formed and solved.

The mass balances consider the air movement in and out of zones, as depicted in Figure 2.1. Each type of mass transfer process depicted in this figure involves the movement of moist air (the mixture of dry air and water vapour). Following the principles of psychrometrics, most BPS tools form separate mass balances for the dry air and water vapour fractions.

2.4 ZONE MASS BALANCE ON DRY AIR

Figure 2.4 represents the zone under consideration and its bounding constructions. A control volume is drawn to delineate the boundary between the zone and its surroundings. Recall from Section 2.2 that this includes the air contained within the building volumes represented by the zone, but may also include furniture, partition walls, and other solid objects. The mass flows crossing the control volume's boundary are also indicated in the figure. This includes sources of moisture, as these sources (e.g. respiration from people) are considered to be situated outside of the zone's control volume.

The mass balance on the dry air contained within the zone can

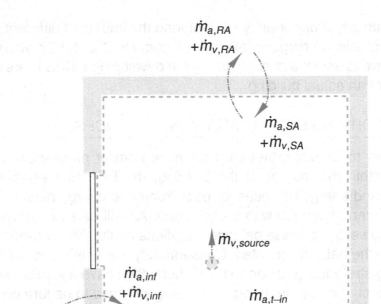

Figure 2.4: Mass transfers into and out of zone control volume (indicated by -----)

be expressed as:

$$\frac{dm_a}{dt} = \sum_{in} \dot{m}_{a,in} - \sum_{out} \dot{m}_{a,out} \qquad (2.1)$$

The terms on the right represent the summation of the dry-air mass flow rates entering and exiting the zone, respectively. The term on the left represents the rate of change of the mass of dry air contained within the zone (m_a). Although changes in pressure and temperature can affect m_a, this transient term is commonly ignored as it is typically very small.

With this assumption and with the expansion of the summation terms to include the three types of dry-air mass flow processes indicated in Figure 2.4, Equation 2.1 can be expressed as:

$$\sum \dot{m}_{a,inf} + \sum \dot{m}_{a,t-in} + \dot{m}_{a,SA} = \sum \dot{m}_{a,exf} + \sum \dot{m}_{a,t-out} + \dot{m}_{a,RA} \quad (2.2)$$

The terms on the left represent the dry-air mass flows into the control volume, while those on the right the exiting dry-air mass flows.

$\dot{m}_{a,inf}$ represents an infiltration of dry air through the building envelope ($\dot{m}_{a,exf}$ is an exfiltration). This could be through an intended opening such as a window, or through an unintended opening such as an imperfection in the air barrier[i]. The term is represented by a summation because there may be many paths through which dry air can flow and each significant one must be represented.

$\dot{m}_{a,t-in}$ represents dry air that is transferred from another zone, while $\dot{m}_{a,t-out}$ represents dry air transferred from the zone under consideration towards another zone in the building. These terms are summed as the zone under consideration may be transferring air with numerous other zones.

$\dot{m}_{a,SA}$ is the dry air supplied by an air-based HVAC system, while $\dot{m}_{a,RA}$ is the dry air extracted from the zone and returned to the HVAC system.

As explained in Section 2.2, it is common to employ the well-mixed assumption. With this, all the air flows leaving the zone are at the same temperature. Consequently, it is convenient to group all of the exiting mass flow rates on the right side of Equation 2.2 into the single term $\dot{m}_{a,out}$ for reasons that will become apparent later:

$$\sum \dot{m}_{a,inf} + \sum \dot{m}_{a,t-in} + \dot{m}_{a,SA} = \dot{m}_{a,out} \quad (2.3)$$

2.5   ZONE MASS BALANCE ON WATER VAPOUR

In a similar manner, a mass balance can be formed on the water vapour contained within zone that is analogous to Equation 2.3:

$$\sum \dot{m}_{v,inf} + \sum \dot{m}_{v,t-in} + \dot{m}_{v,SA} + \sum \dot{m}_{v,source} = \dot{m}_{v,out} + \frac{dm_v}{dt} \quad (2.4)$$

---

[i]Airflow through intentional openings in the building envelope is commonly called *natural ventilation*, and that through unintended openings *infiltration*. These have the same thermal impact and are collectively referred to here as infiltration.

The terms have similar meaning as in Equation 2.3, but in this case the $_v$ subscript indicates water vapour.

In this case the transient term dm_v/dt is not negligible because the zone could be undergoing a humidification or dehumidification process, or the furniture, partition walls, and other solid objects contained within the zone could be storing or releasing moisture. Equation 2.4 also includes the $\dot{m}_{v,source}$ term that represents the sources of water vapour depicted in Figure 2.4. This could be due to moisture gains from occupants, cooking, moisture transfer through the building envelope, etc.

The flow rates of water vapour and dry air can be related to the stream's humidity ratio (ω in kg_v/kg_a), for example:

$$\omega_{inf} = \frac{\dot{m}_{v,inf}}{\dot{m}_{a,inf}} \tag{2.5}$$

Introducing Equation 2.5 to Equation 2.4 explicitly couples the dry-air and water-vapour mass balances:

$$\sum (\omega \cdot \dot{m}_a)_{inf} + \sum (\omega \cdot \dot{m}_a)_{t-in} + (\omega \cdot \dot{m}_a)_{SA} + \sum \dot{m}_{v,source}$$
$$= (\omega \cdot \dot{m}_a)_{out} + \frac{dm_v}{dt} \tag{2.6}$$

2.6 ZONE ENERGY BALANCE

An energy balance in the form of the first law of thermodynamics is also formed for the zone's control volume. As described in Section 2.2, a zone may contain furniture, partition walls, and other solid objects. However, most BPS tools treat the zone control volume as containing only air and provide optional facilities for considering the thermal impact of these solid objects.

The zone energy balance must consider all of the mass transfers considered in the previous sections, because each of these carries energy into or out of the control volume. This energy balance must also consider all of the heat transfer processes that transfer energy between the zone and its surroundings. Only a subset of those shown in Figure 2.1 fall into this category. For example, the solar radiation that is transmitted through the window and absorbed on the internal surface of the floor does not directly cause a transfer of energy between the zone and its surroundings because the zone is considered to contain only air and it does not absorb the radiation. Rather, the solar radiation will be absorbed on the internal surface of the floor, which is outside of the zone's control volume. Likewise the zone does not participate in the radiation exchange between the internal surface of the floor and the internal surface of the right wall.

Only two of the types of heat transfer processes shown in Figure 2.1 actually transfer energy between the zone and its surroundings. These are illustrated in Figure 2.5 along with the mass transfers that cause an energy transfer between the zone and its surroundings.

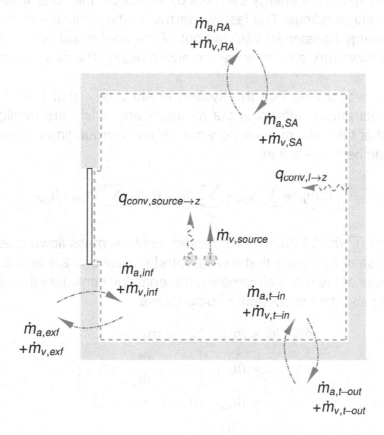

Figure 2.5: Energy balance on zone control volume (indicated by - - - - -)

The general form of the first law of thermodynamics for the zone's control volume can be written as:

$$\frac{d}{dt}\left[m\left(u + \frac{V^2}{2} + gH\right)\right]_z = \sum q_{in} - \sum W_{out}$$

$$+ \sum_{in} \dot{m}_{in} \cdot \left(h + \frac{V^2}{2} + gH\right)_{in} - \sum_{out} \dot{m}_{out} \cdot \left(h + \frac{V^2}{2} + gH\right)_{out} \qquad (2.7)$$

The term on the left represents the transient storage of energy (internal, kinetic, and potential) within the zone control volume (indicated by z). $\sum q_{in}$ is the summation of all of the heat transferred from the surroundings into the control volume, while $\sum W_{out}$ is the summation of the energy transfers by work from the control volume to the surroundings. The last two terms on the right side represent the energy transferred into and out of the control volume with the mass flows (dry air and water vapour) crossing the system boundary.

By recognizing that the system is stationary, that kinetic and potential energy effects at the air inlets and exists are negligible, and that the system does no work on the surroundings, Equation 2.7 can be simplified to:

$$\frac{d}{dt}[mu]_z = \sum q_{in} + \sum_{in} \dot{m}_{in} \cdot h_{in} - \sum_{out} \dot{m}_{out} \cdot h_{out} \qquad (2.8)$$

The $\sum \dot{m} \cdot h$ terms must consider all of the mass flows crossing the system boundary that were treated in Sections 2.4 and 2.5. In the case of the moist air streams, the enthalpy considers that of the dry-air and the water-vapour components:

$$\dot{m}_{in} \cdot h_{in} = \dot{m}_{a,in} \cdot h_{a,in} + \dot{m}_{v,in} \cdot h_{v,in}$$

$$= \dot{m}_{a,in} \cdot \left[h_{a,in} + \frac{\dot{m}_{v,in}}{\dot{m}_{a,in}} \cdot h_{v,in}\right]$$

$$= \dot{m}_{a,in} \cdot \left[h_{a,in} + \omega_{in} \cdot h_{v,in}\right] \qquad (2.9)$$

$$= \dot{m}_{a,in} \cdot h'_{in}$$

where ω_{in} is the humidity ratio (kg_v/kg_a) and h'_{in} is the enthalpy (J/kg_a) of the moist air entering the control volume.

Expanding $\sum q_{in}$ to the heat transfer processes illustrated in Figure 2.5 and considering all the mass transfer processes indicated in that figure, and substituting Equation 2.9 into Equation 2.8 leads to:

$$\frac{d}{dt}[mu]_z = \sum q_{conv,i\rightarrow z} + \sum q_{conv,source\rightarrow z}$$
$$+ \sum \dot{m}_{a,inf} \cdot h'_{inf} + \sum \dot{m}_{a,t-in} \cdot h'_{t-in} + \dot{m}_{a,SA} \cdot h'_{SA} \quad (2.10)$$
$$+ \sum \dot{m}_{v,source} \cdot h_{v,source} - \dot{m}_{a,out} \cdot h'_{out}$$

$q_{conv,i\rightarrow z}$ is the convection heat transfer (W) from surface i to the zone (represented by the z subscript). This term is summed because there is such a heat exchange between the zone and each surface that bounds it (walls, floor, ceiling, windows). $q_{conv,source\rightarrow z}$ represents the convective portion of a heat source (W). This term is summed as there may be multiple sources of heat within the zone, such as lights, computers, appliances, and occupants.

The thermodynamic states in Equation 2.10 are expressed in terms of internal energy and enthalpy. With the aid of some assumptions, this energy balance can be recast to represent the states by temperature. Firstly, the transient term can be reformulated by treating the room air as incompressible with negligible pressure fluctuations:

$$\frac{d}{dt}[mu]_z \approx [m_a c'_P]_z \cdot \frac{dT_z}{dt} \quad (2.11)$$

where T_z is the zone air temperature and c'_P is the specific heat of moist air (J/kg$_a$).

By further approximating c'_P to be constant over the range of temperatures under consideration and by introducing Equation 2.3, the enthalpy terms can be expressed as:

$$\sum \dot{m}_{a,t-in} \cdot \left(h'_{t-in} - h'_{out}\right) \approx \sum (\dot{m}_a \cdot c'_P)_{t-in} \cdot (T_{t-in} - T_{out}) \quad (2.12)$$

Due to its small relative size, it is common for BPS tools to neglect the $\dot{m}_{v,source} \cdot h_{v,source}$ terms in the zone energy balance. Since the zone is treated as being at uniform conditions, the temperature of the air exiting the zone (T_{out}) is equal to the zone air temperature (T_z). This and the incorporation of Equations 2.11 and 2.12 into Equation 2.10 leads to the final form of the zone energy balance:

$$
\begin{aligned}
\left[m_a c_P'\right]_z \cdot \frac{dT_z}{dt} = &\sum q_{conv,i \to z} + \sum q_{conv,source \to z} \\
&+ \sum \left(\dot{m}_a \cdot c_P'\right)_{inf} \cdot (T_{oa} - T_z) \\
&+ \sum \left(\dot{m}_a \cdot c_P'\right)_{t-in} \cdot (T_{t-in} - T_z) \\
&+ \left(\dot{m}_a \cdot c_P'\right)_{SA} \cdot (T_{SA} - T_z)
\end{aligned}
\tag{2.13}
$$

T_{oa} is the temperature of the outdoor air, the temperature at which infiltrating air enters the zone.

With most BPS tools $\left[m_a c_P'\right]_z$ will represent the energy storage of the air contained within the zone. Recall that depending upon the zoning strategy chosen by the user, the zone may also comprise furniture, partition walls, and other solid objects. Most tools provide facilities to allow the user to augment $\left[m_a c_P'\right]_z$ to account for energy storage by these solids, but the responsibility for this is normally left with the user.

2.7 ENERGY BALANCE AT INTERNAL SURFACES

Consider the internal surfaces illustrated in Figure 2.1. Each of these surfaces will experience convection heat transfer with the zone, will exchange radiation in the longwave portion of the infrared spectrum with all of the other internal surfaces it views, and may potentially receive radiation emitted from heat sources or HVAC components. Depending upon the sun position and the surface orientation, solar radiation may be transmitted through windows and be absorbed at the surface. Additionally, there will be heat transfer from the internal surface to the envelope assembly's mass.

All of these heat transfer processes are illustrated schematically for a single internal surface in Figure 2.6, where an infinitesimally thin control volume is drawn around the surface. A first law energy balance in the form of Equation 2.7 can be formed for this control volume. Since the control volume cannot store energy (being infinitesimally thin, it is massless), this results in a balance of the heat

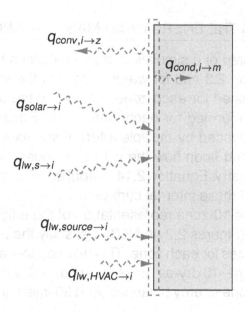

Figure 2.6: Energy balance at internal surface i (control volume indicated by ----)

transfers shown in Figure 2.6:

$$q_{solar \to i} + \sum_s q_{lw,s \to i} + \sum q_{lw,source \to i} + q_{lw,HVAC \to i}$$

$$= q_{conv,i \to z} + q_{cond,i \to m} \tag{2.14}$$

The i subscript represents the internal surface under consideration. $q_{solar \to i}$ is the rate at which incident solar radiation is absorbed by the surface (W).

$q_{lw,s \to i}$ is the net exchange of radiation in the longwave portion of the infrared spectrum from surface s to surface i (W). This term appears in a summation because surface i will exchange radiation with all others surfaces in the zone that it can view.

$q_{cond,i \to m}$ is the heat transfer from surface i into the mass (m) of the materials comprising the envelope assembly (W). The other terms appearing in Equation 2.14 are self-explanatory.

2.8 SOLVING THE ENERGY AND MASS BALANCES

For each timestep of a simulation, mass balances for dry-air in the form of Equation 2.3 and for water vapour in the form of Equation 2.6 are established for each zone defined by the user. An energy balance is also formed for each zone using Equation 2.13. Each zone will be bounded by multiple internal surfaces, the number of which will depend upon how the user has chosen to represent the building's geometry. Equation 2.14 is applied to form an energy balance at each of these internal surfaces.

Consider the 10-zone representation of the office floor plan discussed earlier (Figures 2.2 and 2.3). Let's say the user has defined 9 internal surfaces for each zone. This will lead to—at each timestep of the simulation—10 dry-air mass balances, 10 water-vapour mass balances, 10 zone energy balances, and 90 internal-surface energy balances.

These 120 equations cannot be independently solved because they are highly coupled. For example, the mass flow rate of dry air supplied by the HVAC system to office A ($\dot{m}_{a,SA}$) appears in office A's dry-air and water-vapour mass balances and in office A's zone energy balance: the solution of one of these equations affects the other. And thermal conditions in the reception can influence the functioning of the HVAC system and therefore the magnitude of $\dot{m}_{a,SA}$ flowing to office A.

The energy balances of the surfaces bounding the meeting room zone are coupled through radiation exchange ($q_{lw,s \to i}$). The zone energy balance of office F is coupled to energy balances of office F's bounding internal surfaces through convection heat transfer ($q_{conv,i \to z}$). And the humidity ratio of the corridor appearing in that zone's water-vapour mass balance influences the $c'_{P,t-in}$ of office G's zone energy balance if air flows from the corridor to office G. The couplings between these 120 equations are myriad.

The situation described above is only a partial account of the complexity: Parts III, IV, and V of this book will introduce additional equations that will have to be solved concurrently to these 120. Most BPS tools use one of four techniques to solve this highly coupled (and non-linear) set of equations on a timestep basis to predict the transient thermal performance of the building.

In one approach, the previous timestep's solution is used to initialize an iterative solution for the current timestep. Some terms of the balance equations are linearized (the details of which will be presented in upcoming chapters) and approximated using values from the previous solver iteration. Each of the balance equations is solved independently in turn, producing updated values for each state point which are used in the next solver iteration. The process is repeated until convergence is attained, and then the solver marches forward in time to repeat the entire process for the next simulation timestep.

The second approach avoids iteration within the timestep by directly solving the system of balance equations. This can only be achieved by linearizing the set of equations. The coefficients for some of the terms are evaluated using temperature solutions from the previous timestep. Furthermore, some of the couplings are managed by using state points from the previous timestep. So, for example, if air is flowing from the corridor to office B, the T_{t-in} term in office B's zone energy balance would be taken as the corridor's zone air temperature from the previous timestep. As a result, BPS tools employing this solution approach may be more sensitive to the user's choice of simulation timestep.

The third approach is used only by BPS tools based on the equation-based acausal modelling languages mentioned in the previous chapter (Section 1.5). Rather than applying one of the predefined solution approaches described above, these use symbolic computation (also called computer algebra) to determine appropriate methods to solve the set of governing differential and algebraic equations that define the situation at hand. With this, it is not necessary to linearize the equations or to march through time with a fixed simulation timestep. However, this is often done in practice for the sake of solution stability and computational efficiency (Jorissen *et al.*, 2018).

The fourth technique takes a very different mathematical approach to solving the coupled set of equations. Known as the response factor (or weighting factor) approach, this method was devised to minimize computational burden, which was a key consideration when it was introduced. This method is the subject of the chapter's required reading.

2.9 REQUIRED READING

Reading 2–A

This chapter has described how zone energy and mass balances and energy balances at internal surfaces are formed and solved on a timestep basis in modern BPS tools.

In the early days of BPS such complexity was not possible due to computational limitations. In 1967, Stephenson and Mitalas (1967) introduced the *response factor* method for subdividing the problem domain to minimize the computational burden. Most of the earlier generations of BPS tools—some of which are still in use—were based upon these methods.

Read this article in its entirety and find the answers to the following questions:

1. Why did the authors argue that the response factor method was preferred to finite difference methods for modelling transient heat conduction in building components?
2. What is the principle of superposition? What assumptions must be made to apply this principle?
3. Give an example of an excitation function. Explain how it could be represented by a set of overlapping triangular pulses.
4. What is a room response function? What are some of the parameters that determine a response function?
5. Explain why this method is computationally efficient.
6. Think of a situation where the assumptions imposed by this method may introduce inaccuracies.

2.10 SOURCES FOR FURTHER LEARNING

- Most modern BPS tools form and solve mass and energy balances using techniques like those described in this chapter.

Sowell and Hittle (1995) provide an historical perspective on the contrast between these so-called *heat balance* methods and the response factor (also called weighting factor) methods introduced by Stephenson and Mitalas (1967).

2.11 SIMULATION EXERCISES

These exercises build upon the *Base Case* that was created in Chapter 1. You do not need to modify any inputs for the first three exercises, although you may want to make some refinements and corrections following the previous chapter's exercises. After performing a simulation, you will extract temperature and heat transfer predictions for a single day period—March 6, a cool and sunny day. You will then examine the results of the zone energy balance and of a surface energy balance to reinforce the theory presented in this chapter.

In the last two exercises you will make some geometrical modifications to your *Base Case* to explore the impact of geometrical abstraction and zoning.

Exercise 2–A

Create a temperature-versus-time graph for March 6. Plot the zone air temperature on this graph. Superimpose on this graph the temperature of the internal surfaces of the north wall and of the floor.

Based upon this graph, how do you expect the magnitude and direction of the convection heat transfer between the zone air and the north wall to vary over the day? Do the predicted internal-surface temperatures of the north wall and floor agree with your expectations?

Exercise 2–B

Create a heat transfer-versus-time graph for March 6. Plot the following rates of energy transfer (W) to the zone air (refer to Section 2.6):

- Convection heat transfer from all internal surfaces, $\sum q_{conv,i \to z}$
- Convection heat transfer from sources of heat, $\sum q_{conv,source \to z}$

- Energy transfer due to air infiltration to the zone,

$$\sum \left(\dot{m}_a \cdot c_P' \right)_{inf} \cdot (T_{inf} - T_z)$$

- Energy transfer from the HVAC system,

$$\left(\dot{m}_a \cdot c_P' \right)_{SA} \cdot (T_{SA} - T_z)$$

Observe how the magnitude of the energy transfer due to infiltration varies over the day. Explain the cause of this variation.

Examine your graph and the zone energy balance of Equation 2.13. Although the methods used to calculate convection heat transfer from internal surfaces have not yet been explained (this is the subject of Chapter 4), predict qualitatively how the energy transfer from the HVAC system would be affected if the internal surface convection was increased by 10 %.

Integrate over the day each of the four energy transfer rates listed above and compare their magnitudes (MJ) and directions. Does the sum of these energy transfers equal zero? Provide an explanation by referring to Equation 2.13.

Exercise 2–C

Create another heat transfer-versus-time graph for March 6. Plot the following rates of heat transfer (W) at the internal surface of the north wall (refer to Section 2.7) :

- Absorbed solar radiation, $q_{solar \rightarrow i}$
- Net exchange of radiation in the longwave portion of the infrared spectrum from all other surfaces, $\sum_s q_{lw,s \rightarrow i}$
- Convection to the zone air, $q_{conv,i \rightarrow z}$
- Heat transfer from the surface into the mass of the north wall construction, $q_{cond,i \rightarrow m}$

Examine the direction and magnitude of the heat transfer rates. How does the convection heat transfer rate compare with the explanation you provided in Exercise 2–A?

Although the methods used to calculate the solar radiation absorbed by surfaces have not yet been explained (this will be treated in Chapter 3), based upon your graph and an examination of the energy balance presented in Section 2.7, predict how a 10 % increase in the absorbed solar radiation would impact the other terms in the energy balance.

Exercise 2–D

To explore the impact of geometrical abstraction upon simulation predictions, modify the building's geometry by shortening the length of the building by ~10 % and by extending the depth by ~11 % using the dimensions given in Figure 2.7. This results in no change to the window area, minimal changes to the volume, roof area, and floor area (~0.5 %), and a small change (~3 %) to the opaque wall area.

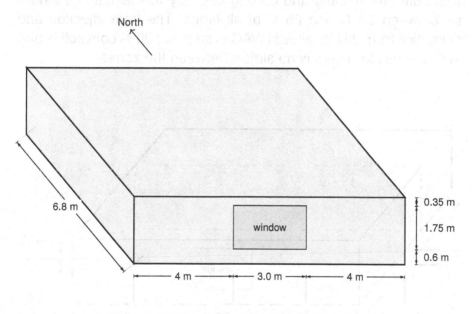

Figure 2.7: Geometry for Exercise 2–D

Perform an annual simulation. What impact does this have on the annual space heating load? And upon the annual space cooling load? Comment on the significance of preserving geometrical details for such an analysis.

Exercise 2–E

Return to the *Base Case* geometry of Chapter 1.

Recall that internal partition walls were ignored and that the building was represented with a single thermal zone. Now modify your geometrical input to represent the building with two thermal zones. One zone will represent a 2 m wide hallway running along the north wall of the building while the second zone will represent the volume adjacent to the south-facing window. Represent the partition wall dividing the zones as a 25 mm thick layer of gypsum using the properties given in Table 1.3. This partition wall is shown in Figure 2.8.

Impose all internal gains to the south-facing zone containing the window. Each zone is conditioned by an idealized HVAC system with sufficient heating and cooling capacity to maintain its indoor air between 20 °C and 25 °C at all times. The heat injection and extraction from this idealized HVAC system is 100 % convective and 100 % sensible. There is no airflow between the zones.

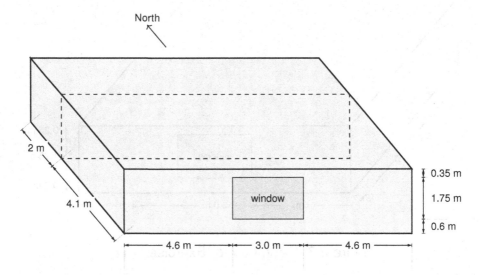

Figure 2.8: Geometry for Exercise 2–E (partition wall between zones indicated by - - - - -)

Perform an annual simulation. Create a temperature-versus-time graph for March 6 and plot the air temperature predicted for each zone on this graph. Is there a temperature difference between the two zones? Why?

Compare this to the graph you created in Exercise 2–A. Comment on the impact that zoning decisions and the inclusion of internal partition walls can have upon the prediction of indoor air temperatures.

Now contrast the annual space heating and space cooling loads with those predicted for the *Base Case* (the one-zone representation). Has subdividing the building into two thermal zones and including the partition wall had a significant impact on these predictions? What additional information was required? In what situations might the additional effort and risk of input errors be justified?

2.12 CLOSING REMARKS

This chapter introduced the concept of the thermal zone and discussed how the user's decisions on zoning and geometrical abstraction have an important impact upon the balance equations that are formed and solved by the BPS tool. It then described how energy and mass balances are formed from a first principles perspective. It was seen that dry-air and moisture mass balances are formed for each thermal zone defined by the user. Energy balances are also formed for each thermal zone, and for each internal building surface bounding each thermal zone. The general approaches in use for solving the resulting highly coupled and non-linear set of equations on a timestep basis were described.

These balance equations contain many terms, each of which describes a significant mass or heat transfer process. Models are used to evaluate each of these terms at each timestep of the simulation. The remaining chapters of Part II of the book describe the models that are used to evaluate the terms for processes occurring within the interior of the building. Chapter 3 explains how the solar radiation absorbed at internal surfaces ($q_{solar \to i}$) is calculated. Chapter 4 shows how the $q_{conv,i \to z}$ term representing convection heat transfer between internal surfaces and the zone is evaluated, while the radiation exchange between internal surfaces in the longwave portion of the infrared spectrum ($q_{lw,s \to i}$) is dealt with in Chapter 5. Chapter 6 discusses how internal sources of heat and moisture are treated: $q_{conv,source \to z}$, $q_{lw,source \to i}$, $\dot{m}_{v,source}$. Finally,

Chapter 7 treats the $\dot{m}_{a,t-in}$ term representing transfer airflow from other zones.

CHAPTER 3

Solar energy absorption by internal surfaces

THIS chapter describes the methods used for calculating the absorption of solar radiation by internal building surfaces. This includes methods used to predict which internal surfaces are irradiated by solar radiation that is transmitted through transparent envelope assemblies.

Chapter learning objectives

1. Understand how solar energy absorption is represented in internal surface energy balances.
2. Realize the complications and necessary approximations involved in estimating the distribution of solar irradiance to internal surfaces.
3. Become aware of tool default methods and their inherent assumptions and limitations.
4. Learn which optional methods are available for more detailed calculations and what additional data must be provided.
5. Develop an appreciation for the impact of prescribed solar absorptivity values on simulation predictions.

3.1 SOLAR PROCESSES RELEVANT TO BUILDINGS

BPS tools must consider many aspects of solar radiation that influence building performance (refer back to Section 1.1). Solar radiation is incident upon the external surfaces of the building envelope (walls, windows, roofs) whenever the sun appears above the horizon. The direction and magnitude of this radiation depends on many factors: scattering by earth's atmosphere, the geometrical relationship between the building and the sun, shading by surrounding objects, reflection off the ground, etc. The methods used to represent these phenomena to predict the solar irradiance upon external surfaces will be treated in Chapter 9.

A transparent envelope assembly (window, skylight) will transmit a portion of the solar radiation incident upon its external surface. The transmitted fraction will depend upon the radiative properties of its glazing layers, incident angle(s), and the presence of shades. These subjects will be treated in Chapter 14. Once transmitted through windows, solar radiation will be reflected and absorbed at internal building surfaces, and a fraction may even be transmitted to another zone or retransmitted back out of the building through windows. This is the topic of the current chapter. We start with a review of some basic solar radiation concepts.

3.2 SOLAR RADIATION BASICS

The sun emits electromagnetic radiation over a wide range of wavelengths (λ). The solar spectrum just outside earth's atmosphere (extraterrestrial) is illustrated in Figure 3.1, which plots spectral irradiance as a function of wavelength. Almost half the extraterrestrial solar radiation lies within the spectrum that is visible to humans ($0.38\,\mu m \lesssim \lambda \lesssim 0.74\,\mu m$), while a considerable amount is at longer (infrared) and to a lesser extent, shorter (ultraviolet) wavelengths.

Each wavelength of radiation reacts differently as it passes through earth's atmosphere. Ozone, oxygen, and nitrogen molecules preferentially absorb the shorter ultraviolet wavelengths, while water vapour and carbon dioxide tend to absorb in the infrared spectrum. Reflection and scattering by nitrogen and oxygen

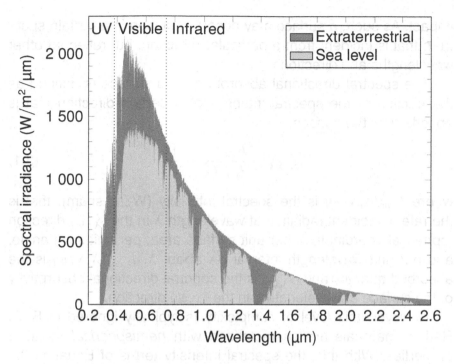

Figure 3.1: Solar spectral irradiance at the top of earth's atmosphere and at sea level [data from ASTM (2014) and ASTM (2012)]

molecules, and by particulates and clouds is also wavelength dependent, although more concentrated in the visible spectrum. As a consequence of these absorptions, reflections, and scatterings, the solar irradiance available on the earth's surface (where buildings are located) is a complex function that depends upon location and prevailing meteorological conditions. Figure 3.1 also plots a typical[i] solar spectrum at sea level, but of course, the actual spectrum that irradiates a building surface will vary from location to location, and from day to day.

Materials that make up building surfaces (brick, painted gypsum, ceramic tiles, etc.) react differently to different wavelengths of radiation. Directional effects (angle of incidence) can also have an

---

[i]The data are taken from ASTM (2012) which provides a standard terrestrial solar spectral irradiance for conditions that are typical of the continental USA.

impact. As such, a surface may absorb radiation in a certain spectrum that is incident from a particular direction, but reflect at other wavelengths and directions.

The spectral directional absorptivity of a surface is defined as the fraction of the spectral intensity of a certain direction that is absorbed by the surface:

$$\alpha_{\lambda,\chi,\psi} = \frac{I_{\lambda,abs}(\lambda, \chi, \psi)}{I_{\lambda,inc}(\lambda, \chi, \psi)} \tag{3.1}$$

where $I_{\lambda,inc}(\lambda, \chi, \psi)$ is the spectral intensity (W/m^2 sr µm), that is the rate of incident radiation at wavelength λ in the (χ, ψ) direction (spherical coordinates), per unit surface area, per unit solid angle, and per unit wavelength interval $d\lambda$ about λ. $I_{\lambda,abs}(\lambda, \chi, \psi)$ is the absorbed quantity and $\alpha_{\lambda,\chi,\psi}$ is the spectral directional absorptivity of the surface at wavelength λ in the (χ, ψ) direction.

However, this level of complexity is typically ignored in BPS. Rather, materials are characterized with *hemispherical* radiative properties. With this, the spectral intensity terms of Equation 3.1 are integrated over a hypothetical hemisphere covering the surface in order to consider radiation from all directions:

$$\alpha_\lambda = \frac{G_{\lambda,abs}(\lambda)}{G_\lambda(\lambda)} \tag{3.2}$$

$G_\lambda(\lambda)$, the spectral irradiance (W/m^2 µm), is the rate at which radiation of wavelength λ is incident upon the surface, per unit surface area, and per unit wavelength interval $d\lambda$ about λ. It is the result of integrating $I_{\lambda,inc}(\lambda, \chi, \psi)$ over the hemisphere. $G_{\lambda,abs}$ is the absorbed quantity, and α_λ is the spectral hemispherical absorptivity.

Although the absorptivity of a surface is a function of wavelength (as implied by Equation 3.2), BPS tools operate with *total* quantities to simplify the analysis. This is accomplished by integrating the spectral irradiance terms over all wavelengths of the solar spectrum $(0.3\,\mu m \lesssim \lambda \lesssim 2.4\,\mu m)$:

$$\begin{aligned}\alpha_{solar} &= \frac{\int_\lambda G_{\lambda,abs}(\lambda)d\lambda}{\int_\lambda G_\lambda(\lambda)d\lambda} \\ &= \frac{G_{solar,abs}}{G_{solar}}\end{aligned} \tag{3.3}$$

where α_{solar} is the total hemispherical absorptivity of the surface in the solar spectrum.

α_{solar} is considered a *total* quantity because it has been integrated over all wavelengths of the solar spectrum, and *hemispherical* because the integration is carried out over all angles of incidence. G_{solar} is the solar irradiance (W/m^2) on the surface coming from all directions and at all wavelengths of the solar spectrum, while $G_{solar,abs}$ is the quantity of this irradiance that is absorbed by the surface.

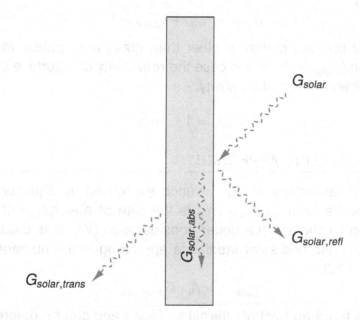

Figure 3.2: Absorption, transmission, and reflection of incident irradiance

The solar irradiance incident upon a surface will be absorbed, transmitted, or reflected (see Figure 3.2):

$$G_{solar} = G_{solar,abs} + G_{solar,trans} + G_{solar,refl} \qquad (3.4)$$

As with absorptivity, the total solar hemispherical transmissivity (τ_{solar}) and the total solar hemispherical reflectivity (ρ_{solar}) are defined to represent the fractions of the solar irradiance that are transmitted and reflected:

$$\tau_{solar} = \frac{G_{solar,trans}}{G_{solar}} \tag{3.5}$$

$$\rho_{solar} = \frac{G_{solar,refl}}{G_{solar}} \tag{3.6}$$

Substitution of Equations 3.3, 3.5, and 3.6 into Equation 3.4 leads to:

$$\alpha_{solar} + \tau_{solar} + \rho_{solar} = 1 \tag{3.7}$$

Most building materials other than glass are opaque to solar radiation ($\tau_{solar} = 0$). In this case the reflectivity of a surface can be determined from its absorptivity:

$$\rho_{solar} = 1 - \alpha_{solar} \tag{3.8}$$

3.3   MODELLING APPROACH

The internal-surface energy balance expressed by Equation 2.14 includes the term $q_{solar \rightarrow i}$. This is the rate of absorption of solar radiation by the surface under consideration (W). It is calculated from the absorbed solar irradiance appearing in the numerator of Equation 3.3:

$$q_{solar \rightarrow i} = A_i \cdot G_{solar,abs \rightarrow i} \tag{3.9}$$

A_i is the area (m^2) of internal surface i and can be determined from the geometrical input provided by the user. The $_{\rightarrow i}$ suffix has been added to make clear that the irradiance term pertains to the surface under consideration.

Recall that an energy balance is established and solved for each internal surface of each zone. As such, for each timestep of the simulation, unique values of $q_{solar \rightarrow i}$ must be determined for each internal surface of the building.

Substituting Equation 3.3 into Equation 3.9 leads to an expression for the desired rate of heat transfer to surface i at a given timestep in terms of solar absorptivity and incident irradiance:

$$q_{solar \rightarrow i} = \alpha_{solar,i} \cdot A_i \cdot G_{solar \rightarrow i} \tag{3.10}$$

$G_{solar \to i}$ is the solar irradiance (W/m²) incident upon the surface. A portion of this irradiance will be *beam* (sometimes called *direct*) and a portion will be *diffuse* (see Figure 3.3). Beam radiation has travelled from the sun without being scattered by earth's atmosphere or by reflecting (e.g. ground) and transmitting (e.g. glazing) surfaces. As such, beam irradiance has a single direction and will strike surfaces preferentially depending upon this direction and the room's geometry. In contrast, diffuse irradiance has been scattered by earth's atmosphere or by reflecting or transmitting surfaces, and therefore has many directions.

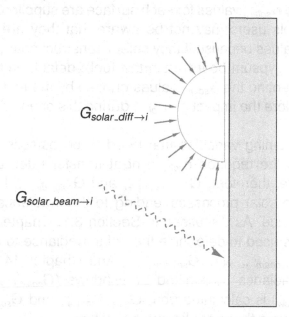

$G_{solar\_diff \to i}$

$G_{solar\_beam \to i}$

Figure 3.3: Beam and diffuse irradiance of an internal surface

Because the surface is characterized with a total hemispherical absorptivity, the solar irradiance term required by Equation 3.10 is simply a sum of the beam and diffuse components:

$$G_{solar \to i} = G_{solar\_beam \to i} + G_{solar\_diff \to i} \qquad (3.11)$$

Substituting Equation 3.11 into Equation 3.10 yields:

$$q_{solar \to i} = \alpha_{solar,i} \cdot A_i \cdot (G_{solar\_beam \to i} + G_{solar\_diff \to i}) \qquad (3.12)$$

This equation implicitly assumes that both the beam and diffuse irradiance is uniform over the surface. This may not be strictly correct. For example, at high solar elevations (when the sun is high in the sky) the bottom portion of a wall may receive beam irradiance, but not the top, depending upon the geometrical relationships between the surface, the window, and the sun. However, this assumption is necessary given that the energy balance of Equation 2.14 is formed for the internal surface in its entirety[ii].

The A_i and $\alpha_{solar,i}$ values required by Equation 3.12 are time-invariant. A_i is determined from the geometrical input provided by the user and $\alpha_{solar,i}$ values for each surface are supplied by the user. In some tools users may not be aware that they are prescribing the $\alpha_{solar,i}$ values because if they select construction materials (e.g. tile, painted gypsum board) from the tool's default databases they may be accepting the $\alpha_{solar,i}$ values chosen by the tool developers. You will explore the impact of $\alpha_{solar,i}$ during this chapter's simulation exercises.

The remaining variables that need to be established in order to calculate the required $q_{solar \rightarrow i}$ heat transfer rates using Equation 3.12 are, therefore, $G_{solar\_beam \rightarrow i}$ and $G_{solar\_diff \rightarrow i}$. Figure 3.4 illustrates the solar processes leading to the irradiance of a wall's internal surface. As discussed in Section 3.1, Chapter 9 will treat the methods used to determine the solar irradiance to exterior surfaces: $G_{solar\_beam \rightarrow e}$ and $G_{solar\_diff \rightarrow e}$. And Chapter 14 will explain how the irradiance transmitted by windows ($G_{solar\_beam,window}$ and $G_{solar\_diff,window}$) is calculated from $G_{solar\_beam \rightarrow e}$ and $G_{solar\_diff \rightarrow e}$. For now we assume these are known quantities.

The next two sections describe the methods that are employed by BPS tools for calculating $G_{solar\_beam \rightarrow i}$ and $G_{solar\_diff \rightarrow i}$ from $G_{solar\_beam,window}$ and $G_{solar\_diff,window}$.

3.4 DISTRIBUTION OF SOLAR BEAM IRRADIANCE

The distribution of the beam irradiance transmitted through a window to a zone's internal surfaces will depend upon sun position,

---

[ii]The potential limitations of this assumption can be overcome if the user subdivides the wall into numerous segments, a practice allowed by many BPS tools.

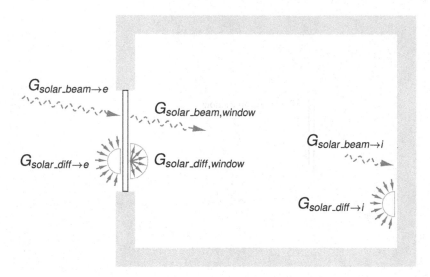

Figure 3.4: Solar processes leading to irradiance of a wall's internal surface

window location, and room geometry. Figure 3.5 illustrates a situation where a portion of $G_{solar\_beam,window}$ is incident upon a wall and a portion upon the floor.

The fraction of the transmitted beam radiation that is incident upon surface i can be represented by a scalar that is sometimes called an *insolation factor*:

$$\zeta_i = \frac{A_i \cdot G_{solar\_beam \to i}}{A_{window} \cdot G_{solar\_beam,window}} \tag{3.13}$$

From a first law energy balance it can be seen that the transmitted beam radiation incident over all surfaces must equal the energy transmitted through the window:

$$A_{window} \cdot G_{solar\_beam,window} = \sum_i A_i \cdot G_{solar\_beam \to i} \tag{3.14}$$

Combining Equations 3.13 and 3.14 shows that a zone's insolation factors must sum to unity:

$$\sum_i \zeta_i = 1 \tag{3.15}$$

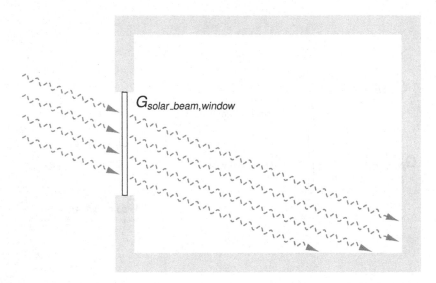

Figure 3.5: Distribution of solar beam irradiance to internal surfaces

Therefore, once ζ_i is established the value of $G_{solar\_beam \rightarrow i}$ required by Equation 3.12 can be determined from Equation 3.13 from the geometry and $G_{solar\_beam, window}$.

In reality, each internal surface of a zone will have a unique ζ_i and these will vary with each timestep of the simulation, depending upon sun position, window location, and room geometry. Various methods are employed by BPS tools to determine these values.

The most rigorous approach is to consider the geometrical relationships between the sun and each window, and between each window and each internal surface. As the sun (apparently) moves through the sky during the simulation, calculations can be performed to determine the ζ_i of each surface.

These calculations can be performed with direct geometrical methods that employ analytical solutions, or beam projection methods that involve discretizing windows and internal surfaces into small areas to determine the distribution of a sun patch. Either approach could accurately depict the situation illustrated in Figure 3.5. You will learn more about these detailed calculation methods through this chapter's required reading.

However, users should be aware that rigorous methods such as these are rarely employed by default by BPS tools. This is something of an anachronism because in the past the computational resources required for such calculations were considered to be too expensive for routine application. Invoking these optional methods usually requires the user to make a conscientious choice and to provide some additional (typically minimal) information.

Users should also be aware of the limitations of these methods. For example, direct geometrical methods can only work for particular zone geometries. The algorithms may not work, for example, for a zone that has an L-shaped or T-shaped floor plan. And for the sake of computational efficiency, the calculations are not usually performed for every timestep of the simulation. One common approach is to perform the calculations only for the middle day of the month and to assume that the sun path changes minimally over the month.

The default treatments employed by BPS tools are far less rigorous than the detailed methods described above. One common method is to assume uniform beam irradiance to all internal surfaces. This is equivalent to assuming that the beam irradiance is distributed to all internal surfaces based on area weighting:

$$\zeta_i = \frac{A_i}{\sum_k A_k} \qquad (3.16)$$

where the summation occurs over all the internal surfaces (denoted by k) of the zone.

With this approach, the energy balance for the ceiling of Figure 3.5 would receive the same quantity of $q_{solar\_beam \rightarrow i}$ as that for the floor (if they have the same α_{solar}). Also, the wall containing the window would receive some of the beam irradiance transmitted by the window, something that would not occur in reality.

Another commonly used method is to assume that all of the beam radiation is incident upon the floor. In this case $\zeta_i = 1$ for the floor and $\zeta_i = 0$ for all other surfaces. With this approach the beam irradiance to the wall in Figure 3.5 would not be considered in that surface's energy balance, but would instead be added to the energy balance for the floor.

Clearly, the artefacts of these simplified modelling approaches will distort the internal surface energy balances. During this chapter's simulation exercises you will explore which method your chosen BPS tool uses by default and which optional methods it supports.

3.5 DISTRIBUTION OF DIFFUSE AND REFLECTED SOLAR IRRADIANCE

In reality, the diffuse irradiance transmitted through a window may have different intensities in different directions. However, a common approximation made by BPS tools is to treat this irradiance as equally intense in all directions. This is known as hemispherically diffuse radiation. With this assumption, $G_{solar\_diff,window}$ is distributed to a zone's internal surfaces based on area weighting:

$$Z_i = \frac{A_i}{\sum_k A_k} \qquad (3.17)$$

Where Z_i is the insolation factor for diffuse irradiance and is analogous to the insolation factor for beam irradiance:

$$Z_i = \frac{A_i \cdot G_{solar\_diff \rightarrow i}}{A_{window} \cdot G_{solar\_diff,window}} \qquad (3.18)$$

A variant on this approach which is also commonly employed limits the summation in the denominator of Equation 3.17 to non-coplanar surfaces. With this, the wall containing the window in Figure 3.5 would not receive diffuse irradiance transmitted through the window.

A portion of both the beam and diffuse irradiance incident upon opaque surfaces will be reflected according to Equation 3.8. Most commonly these reflections are treated as diffuse and are distributed to other surfaces using the Z_i values as discussed above.

3.6 REQUIRED READING

Reading 3–A

Chan and Tzempelikos (2013) explain that most BPS tools make simplifications in calculating the distribution of solar irradiance to internal surfaces. This reduces the computational burden of simulations and minimizes the geometrical details required from users. They describe the various methods in use for treating beam and diffuse components and summarize which methods are available in some common BPS tools. They also investigate the accuracy of the various methods by comparing against a detailed calculation model that has been empirically validated.

Read this article in its entirety and find answers to the following:

1. Do the simplified solar-beam distribution models lead to greater errors when the windows are large or small? Explain why.

2. Do the simplified solar-beam distribution models lead to greater or smaller errors when the floor and ceiling have more mass? Explain why.

3. Explain why Case 3 ($\zeta_i = 1$ for the floor) results in a larger error in predicting heating than in predicting cooling.

4. Are simulation predictions more sensitive to the modelling of $G_{solar\_beam \rightarrow i}$ or $G_{solar\_diff \rightarrow i}$?

5. What magnitude of errors should the user expect when using a BPS tool that employs simplified models to predict the distribution of solar irradiance to internal surfaces?

3.7 SOURCES FOR FURTHER LEARNING

- Chatziangelidis and Bouris (2009) discuss the importance of accurately calculating the distribution of solar energy entering buildings. The introduction to this paper describes approaches of varying complexity, accuracy, and computational requirements.

- Hiller *et al.* (2000) describe the TRNSHD program (part of TRNSYS) for external shading and insolation calculations. It includes an empirical validation of the program's determination of the internal solar distribution of beam radiation using measurements taken in a rectangular room configured with two different windows.

- Chapter 7 of Clarke (2001) details ESP-r's beam projection method for calculating insolation patterns.

3.8 SIMULATION EXERCISES

Revert your BPS tool inputs to represent once again the *Base Case* described in Section 1.9, including any refinements or corrections you made following the previous exercises. Perform an annual simulation to produce a fresh set of *Base Case* results for use in the following exercises.

Exercise 3–A

Solar absorptivity values were specified for the materials forming the internal surfaces of all the opaque envelope constructions in the *Base Case*. Now increase the concrete floor's α_{solar} from 0.6 to 0.9 and perform another annual simulation using the same timestep.

What impact does this change have upon the annual space heating load? And upon the annual space cooling load? Are these results in line with your expectations?

What are possible sources of information that can be used to determine α_{solar} values for materials?

Exercise 3–B

Extract the results for March 6 for both the *Base Case* and for the simulation you conducted in Exercise 3–A. Create a graph that plots $q_{solar \rightarrow i}$ for the concrete floor versus time for this day. Did increasing the concrete's α_{solar} have the expected impact upon $q_{solar \rightarrow i}$ for the floor? (Refer to Equation 3.12.)

Create a second graph for plotting the rate of heat input from the HVAC system versus time and contrast the results from the two simulations. Explain why this significant change in the concrete's α_{solar} had minimal impact upon heating.

Exercise 3–C

Revert to the *Base Case* (i.e. reset the concrete's α_{solar} to 0.6).

How did you treat the distribution of solar gains to internal surfaces? If you did not use the default method provided by your BPS tool, then configure your tool now to employ its default method. Review your BPS tool's documentation to determine its default method. Relate this to the theory presented in Sections 3.4 and 3.5 and Reading 3–A.

Does your BPS tool offer optional methods for calculating insolation? Describe the optional methods that are available. What additional data must the user provide? Now configure your BPS tool to apply its most detailed method and perform another simulation.

What impact does this change have upon the annual space heating load? And upon the annual space cooling load?

Exercise 3–D

Extract the results for March 6 for both the *Base Case* and for the simulation you conducted in Exercise 3–C. Create a graph that plots $q_{solar \rightarrow i}$ for the concrete floor versus time for this day.

Are these results consistent with your understanding of your BPS tool's default and optional methods? Describe a situation in which the default method might result in significant prediction errors.

3.9 CLOSING REMARKS

This chapter explained that BPS tools characterize materials using total hemispherical radiation properties in the solar spectrum, and that users must prescribe the solar absorptivity of all internal surfaces (or rely on tool default values).

Through the required reading and the theory presented in this chapter, you learnt about the types of methods that are available for predicting the distribution of solar irradiance to internal surfaces. Some are detailed and some are quite simplistic. You became aware of which methods are available in your chosen BPS tool through the simulation exercises, and realized the actions required to invoke them.

CHAPTER 4

Convective heat transfer at internal surfaces

THIS chapter describes the methods used for treating convection heat transfer at internal building surfaces.

Chapter learning objectives

1. Understand how convection heat transfer is represented in zone energy balances and internal surface energy balances.
2. Appreciate the complexity and uncertainty in determining convection coefficients.
3. Realize the sensitivity of simulation predictions to tool defaults and user choices.
4. Understand the options available for treating internal surface convection and the level of effort and data each approach requires.

4.1 MODELLING APPROACH

Chapter 2 showed how energy balances are formed on the zone (Equation 2.13) and at internal surfaces (Equation 2.14). Convection heat transfer between internal surfaces and the zone ($q_{conv,i \to z}$) appeared in both of these equations.

Chapter 2 also explained that it is common for BPS tools to employ the well-mixed assumption, treating the thermal zone as having uniform conditions (e.g. temperature, humidity). This assumption allows the internal surface convection to be modelled using Newton's law of cooling because T_z represents the temperature of the entire zone:

$$q_{conv,i \to z} = A_i \cdot h_{conv,i} \cdot (T_i - T_z) \qquad (4.1)$$

A_i is the area (m²) of internal surface i and can be determined from the geometrical input provided by the user. $h_{conv,i}$ is the convection coefficient (W/m² K), which essentially characterizes the convection regime. In reality, its value depends upon the local flow regime and will vary from surface to surface, and in time. The crux of modelling internal surface convection is therefore establishing appropriate $h_{conv,i}$ values.

4.2 CONVECTION HEAT TRANSFER BASICS

In the heat transfer literature, the convection coefficient is typically represented using the dimensionless Nusselt number (Nu), which represents the ratio of convective to conductive heat transfer:

$$Nu = \frac{h_{conv} L}{k} \qquad (4.2)$$

where L is a length scale (m), perhaps representing the length of a flat plate, the diameter of a sphere, or the height of a wall. k is the thermal conductivity (W/m K) of the fluid involved in the convection heat transfer.

Experiments have been conducted for various geometrical configurations to establish empirical relationships between Nu and other dimensionless quantities that characterize the flow regime. For forced flow situations the flow regime is typically characterized by the Reynolds number, which represents a ratio between momentum and viscous effects:

$$Re = \frac{\rho VL}{\mu} \tag{4.3}$$

where ρ is the fluid's density (kg/m³) and μ its viscosity (N s/m²). V is the velocity (m/s) of the fluid which is forced to flow over the surface.

Any basic heat transfer textbook contains many correlations for forced flow situations, for example (Incropera *et al.*, 2007):

$$Nu = C \cdot Re^m \cdot Pr^{1/3} \tag{4.4}$$

Pr is the Prandtl number, a dimensionless quantity representing the ratio of momentum to thermal diffusivity:

$$Pr = \frac{c_P \mu}{k} \tag{4.5}$$

where c_p is the specific heat of the fluid (J/kg K).

The C and m values in Equation 4.4 are empirical constants that are determined by fitting measured data to the functional form. Values depend upon the shape of the surface and the Re number and are tabulated for various scenarios. However, most of these situations—cross flow over a hexagonal cylinder, forced flow over an infinitely long plate, etc.—are not representative of surfaces found within buildings.

For buoyancy-driven flows—known as natural convection—the Rayleigh number (Ra), representing the ratio of buoyant forces to viscous forces, is commonly used to characterize the flow regime:

$$Ra = \frac{g\beta_v c_P \rho^2 |T_z - T_i| L^3}{\mu k} \tag{4.6}$$

where g is gravitational acceleration (m/s²) and β_v is the fluid's volumetric thermal expansion coefficient (1/K).

Heat transfer textbooks also contain a multitude of correlations for natural convection situations, such as (Incropera *et al.*, 2007):

$$Nu = C \cdot Ra^m \tag{4.7}$$

The C and m values in Equation 4.7 are once again empirical constants that are determined by fitting measured data to the functional form. Although many tabulated correlations are available in the heat transfer literature, most—spheres, cylinders of various cross-sectional shapes, etc.—do not correspond well to situations found within buildings.

4.3 CONVECTIVE REGIMES WITHIN BUILDINGS

It would be an impossible task to develop and implement models for predicting convection coefficients for all possible flow regimes encountered within buildings. Even the presence and location of furniture, and the movement and metabolic functioning of occupants alter indoor airflow patterns, and thus convection heat transfer at internal surfaces. The pragmatic approach that has been commonly adopted is to broadly classify the airflows encountered within buildings and to select appropriate correlations for each case.

The forces that drive indoor airflow can be described as either mechanical or buoyant. Mechanical forces are generally caused by fans or by wind entering through openings. Fans can be located within the room or within air-based HVAC systems that supply heated or cooled air to the space. Buoyant forces can result from heat sources located within the room (radiators, occupants, office equipment, etc.) or from surface-to-air temperature differences. The surface-to-air temperature differences can be caused by heat transfer through the building envelope (e.g. the cold surface of a window), solar insolation, or fabric-embedded conditioning devices (e.g. in-floor heating, chilled ceiling panels). In some cases, both mechanical and buoyant forces can be significant drivers of room air motion.

Further subdivision is possible. For example, in the case of buoyancy-driven flow resulting from the presence of a wood stove, whether the stove is located in the middle of the room, or next to an external window, will have an effect on the room's airflow pattern. In the latter case, the warm plumes rising from the stove will be cooled by heat transfer through the window, causing a competition in the buoyant effects. Similarly, in the case of mechanically driven

Table 4.1: Classification of indoor convection regimes

Nature of flow	Driving force
Buoyant	Hot/cold internal surfaces
	Solar insolation to internal surfaces
	Radiant floor heating/chilled beams
	Terminal heating/cooling devices
Forced	Supply of ventilation air
	Air-based HVAC supplies/extracts
	Fan-assisted terminal heating/cooling devices
Mixed	Combination of buoyant and forced

flows, the location of the supply air diffuser and the extract will influence which walls experience wall jet flow and which experience impinging flow. Clearly some factors have a greater influence than others. Whether the room is mechanically ventilated or not has a more profound influence on the convective regime than does the location of diffusers and extracts.

The convective regimes encountered within buildings can be broadly classified as in Table 4.1.

4.4   CONVECTION CORRELATIONS FOR BUILDINGS

Correlations have been developed to characterize many of the flow regimes outlined in Section 4.3. Alamdari and Hammond (1983) were one of the first to develop convection correlations specific to buildings. Rather than conducting new experiments, they drew upon data reported in the literature to develop their correlations. They provide separate correlations for: vertical surfaces; stably-stratified horizontal surfaces (e.g. warm air above a cool floor); and buoyant flow from horizontal surfaces (e.g. cool air above a warm floor).

Their correlations span the full range of temperatures and dimensions relevant to building applications, and are cast in dimensional form rather than the dimensionless form preferred in the heat transfer literature (e.g. Equations 4.4 and 4.7) to simplify their implementation into BPS tools. For example, Alamdari and Hammond provide this correlation for vertical surfaces:

$$h_{conv,i} = \left\{ \left[1.5 \cdot \left(\frac{|T_z - T_i|}{H} \right)^{1/4} \right]^6 + \left[1.23 \cdot \left| T_z - T_i \right|^{1/3} \right]^6 \right\}^{1/6}$$

(4.8)

where H is the height of the vertical surface (m).

Others (e.g. Khalifa and Marshall, 1990; Awbi and Hatton, 1999; Fisher and Pedersen, 1997) conducted experiments in room-sized test cells to produce correlations specific to internal convection within buildings. For example, Khalifa and Marshall varied the configuration of their test cell and repeated experiments to assess a number of common convection regimes. The measured data were used to derive quantities of interest, and these data regressed to form new correlations for specific situations. For example, they developed the following correlation for the convection coefficient at window surfaces for rooms heated by radiators situated underneath windows:

$$h_{conv,i} = 8.07 \cdot \left| T_z - T_i \right|^{0.11}$$

(4.9)

Many other correlations for $h_{conv,i}$ exist, as you will discover during this chapter's required readings.

4.5 COMMON APPROACHES FOR DETERMINING CONVECTION COEFFICIENTS

Recall from Section 4.1 that in reality $h_{conv,i}$ values vary from surface to surface, and with time. There are four common approaches used in BPS for establishing their values:

1. Time-invariant values are used, perhaps unique values for each surface. These are either user-prescribed or defaulted by the BPS tool. In such cases the user should be aware of the implications this may have upon the energy balances.
2. The BPS tool's default correlation is used to recalculate $h_{conv,i}$ values each timestep of the simulation. The tool may assign a default correlation based upon the surface's orientation (e.g. vertical).

3. A user-selected correlation is chosen for each surface and $h_{conv,i}$ values are recalculated each timestep.

4. $h_{conv,i}$ values are recalculated each timestep using a correlation selected by the BPS tool wherein this selection adapts to the simulated conditions. For example, a forced flow correlation may be used when an air-based HVAC system operates and a buoyancy-driven correlation may be used when the HVAC system is inoperative.

You will explore which of these methods are supported by your chosen BPS tool during this chapter's simulation exercises.

4.6   REQUIRED READING

Reading 4–A

This chapter has described the importance of using appropriate correlations to calculate $h_{conv,i}$ coefficients for the flow regimes at hand. Peeters *et al.* (2011) summarize the available correlations for natural, forced, and mixed convection that are appropriate for BPS and describe how each was derived.

Read Section 2 of this article in detail and find answers to the following questions:

1. What are the independent variables that are used as inputs to correlations for natural convection?

2. What are the independent variables that are used as inputs to correlations for forced convection?

3. What major assumption is implied by using similarity-based correlations?

4. This chapter explained that it is common for BPS tools to apply the well-mixed assumption with Newton's law of cooling (Equation 4.1). What is a major complication in using some of the available natural convection correlations with this approach? (Hint: see "reference temperature".)

5. What are some of the major sources of error in the experiments that have been performed to derive correlations for natural and forced flows?

6. Describe a situation for which both natural and forced flow conditions will occur.

Reading 4–B

Beausoleil-Morrison (2002) discusses how BPS tools can adapt the calculation of convection coefficients during the course of a simulation.

Read this article and find answers to the following questions:

1. Describe a situation that would warrant the complexity required by this level of modelling resolution.

2. What additional data are required from the user?

3. What level of uncertainty is introduced when time-invariant convection coefficients or correlations are employed in a simulation?

4.7   SOURCES FOR FURTHER LEARNING

- Section 3 of Peeters *et al.* (2011) examines the implications of using the well-mixed assumption, the choice of length scales, and the impact of obstructions on convective regimes.

4.8   SIMULATION EXERCISES

Revert your BPS tool inputs to represent once again the *Base Case* described in Section 1.9, including any refinements or corrections you made following the previous exercises. Perform an annual simulation to produce a fresh set of *Base Case* results for use in the following exercises.

Exercise 4–A

How did you treat internal surface convection in your *Base Case*? Did you use your BPS tool's default approach? This is likely the case if you did not take action to override your tool's default method. Consult your BPS tool's help file or technical documentation to determine its default approach for establishing $h_{conv,i}$ coefficients.

If you used your BPS tool's default method, then configure your tool to now impose time-invariant $h_{conv,i}$ coefficients of $3\,W/m^2\,K$ at all internal surfaces and perform another annual simulation using the same timestep. This is the first approach listed in Section 4.5.

If you did not use the default method provided by your BPS tool for the *Base Case*, then configure your tool to now employ its default method and perform another annual simulation using the same timestep. This is the second approach listed in Section 4.5.

Compare the results of the two simulations. What impact does this change have upon the annual space heating load? And upon the annual space cooling load?

Exercise 4–B

Extract results for March 6 for both the *Base Case* and the Exercise 4–A simulation.

Create a graph that plots $q_{conv,i \to z}$ for the window from the two simulations. Over the course of this day, how does this rate of heat transfer differ between the two simulations?

Examine the $h_{conv,i}$ coefficients for the simulation in which your BPS tool applies its default method. How does this coefficient vary over the day? How does this explain the differences between the $q_{conv,i \to z}$ predictions illustrated in the graph?

Create a second graph for plotting the rate of heat input from the HVAC system versus time, and contrast the results from the two simulations. Comment on how the user's choice of approach for determining $h_{conv,i}$ coefficients can impact the predicted timing and magnitude of HVAC energy transfers.

Exercise 4–C

Does your BPS tool support the third approach listed in Section 4.5? If so, which of the correlations listed in Reading 4–A does your tool support? Does it support additional correlations as well?

Configure your BPS tool to apply one or more of its optional correlations and perform some additional simulations. What impact do these changes have upon the annual space heating load? And upon the annual space cooling load?

Exercise 4–D

Does your BPS tool support the fourth approach listed in Section 4.5? If so, configure your BPS tool to use this approach. Assume that the heating system is composed of a circulating fan that impinges upon the window and the south wall containing the window.

What impact does this change have upon the annual space heating load? And upon the annual space cooling load?

Add the predictions from this simulation to the $q_{conv,i \to z}$ graph created in Exercise 4–B. How does $q_{conv,i \to z}$ differ between the three simulations when the HVAC system is active? And when the HVAC system is inactive?

Comment on the level of effort required to invoke this more detailed modelling approach and contrast this to your expectations based upon Reading 4–B.

4.9 CLOSING REMARKS

This chapter explained that, thanks to the commonly employed well-mixed assumption, the convection heat transfer between internal surfaces and the zone can be determined using Newton's law of cooling. With this, convection regimes—which in reality will vary from surface-to-surface and in time—are characterized by convection coefficients.

Through the required readings you became aware of the numerous correlations used in the BPS field for calculating these coefficients. You discovered your BPS tool's default and optional methods for treating convection at internal surfaces by conducting the simulation exercises. These also helped you appreciate the impact that user choices (including relying on default methods) can have upon simulation predictions.

CHAPTER 5

Longwave radiation exchange between internal surfaces

THIS chapter describes the options for calculating longwave radiation exchange between internal building surfaces, including methods for determining view factors between these surfaces.

Chapter learning objectives

1. Understand how longwave radiation exchange is represented in internal surface energy balances.
2. Learn how to determine which simplifying assumptions are employed by your chosen BPS tool.
3. Understand tool default and optional methods for determining view factors.
4. Develop an appreciation for the impact of prescribed longwave radiation properties and view factors on simulation predictions.

5.1 RADIATION BETWEEN INTERNAL SURFACES

Each of a zone's internal surfaces will emit radiation, and this radiation will be absorbed and/or reflected by other internal surfaces (see Figure 2.1). The exchange of radiant energy between the internal surfaces will depend upon many factors, including geometrical relationships between surfaces and their radiation properties.

The internal-surface energy balance developed in Chapter 2 (Equation 2.14) included the term $\sum_s q_{lw,s \rightarrow i}$, where $q_{lw,s \rightarrow i}$ is the net rate of radiation exchange from surface s to surface i. The summation is performed over all of the zone's internal surfaces (denoted by s) that can be *viewed* by the surface under consideration, surface i. The subscript $_{lw}$ indicates that this radiation exchange is occurring in the longwave portion of the infrared spectrum.

This chapter describes the methods that are commonly used by BPS tools to calculate the $q_{lw,s \rightarrow i}$ terms. We begin with a review of some basic concepts related to radiation exchange.

5.2 EMISSION OF LONGWAVE RADIATION

All objects continuously emit electromagnetic radiation due to atomic and molecular agitation associated with their internal energy. This type of electromagnetic radiation is known as *thermal radiation*. We already encountered thermal radiation in Section 3.2, which discussed the electromagnetic radiation emitted by the sun. As with the sun, thermal emission from all other objects occurs over a wide range of wavelengths.

The *blackbody* is an idealization of a perfect emitter. The spectral emissive power of a blackbody can be determined with the well-known Planck distribution (refer to any basic heat transfer textbook), which shows that both magnitude and spectral distribution of emission are strong functions of the blackbody's temperature.

Figure 5.1 plots the Planck distribution of blackbodies at three temperatures. The sun emits approximately as a blackbody at 5800 K. As can be seen in the figure, the peak emission of a blackbody at this temperature occurs within the visible spectrum, although there is also significant emission at wavelengths in the

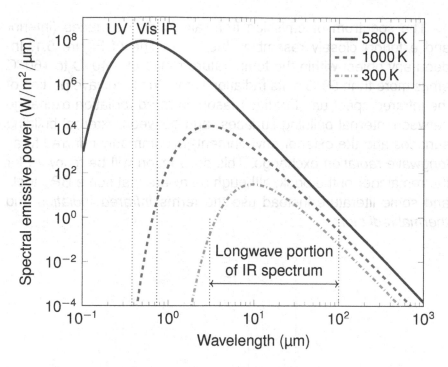

Figure 5.1: Spectral emissive power of blackbodies at various temperatures (log-log scale)

ultraviolet and infrared spectrums. We already encountered this in Section 3.2 in the context of extraterrestrial solar irradiance.

By contrasting the ——— and ----- lines, it can be seen from Figure 5.1 that the peak spectral emissive power from a blackbody at 1000 K is 10 000 times lower than that from a blackbody at 5800 K. (The figure is plotted on a log-log scale.) Also, emission from the cooler blackbody occurs predominantly at longer wavelengths.

Figure 5.1 also plots the emission from a 300 K blackbody. Its emission is entirely within the infrared spectrum, with the peak occurring around 10 μm. Very little of its emission occurs within the shorter wavelengths of the infrared spectrum (sometimes called *near infrared*). Rather, more than 99 % of the emitted energy is within the so-called longwave portion of the infrared spectrum (3 μm $\lesssim \lambda \lesssim$ 100 μm).

The spectrum of emission from all building surfaces (interior and exterior) closely resembles the ------ line of Figure 5.1. Indeed any object within the temperature range of $-40\,°C$ to $+60\,°C$ emits more than 99 % of its radiation within the longwave portion of the infrared spectrum. For this reason, infrared radiation exchange between internal building surfaces (and between external building surfaces and the exterior environment) is commonly referred to as *longwave radiation* exchange. This convention will be followed for the remainder of the book, although be aware that some BPS tools and some literature instead use the terms *infrared radiation* and *thermal radiation*.

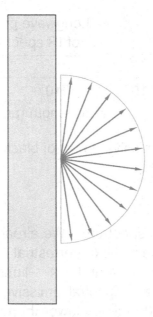

Figure 5.2: Diffuse emission

As blackbodies emit equally in all directions (see Figure 5.2) they are called diffuse emitters. The total emissive power of a blackbody, that is the radiant flux emitted over all wavelengths and over a hypothetical hemisphere covering the surface, is given by the well-known Stefan-Boltzmann law:

$$E_b = \sigma \cdot T^4 \tag{5.1}$$

where T is the blackbody's temperature (K) and E_b is its emissive power (W/m^2). σ is the Stefan-Boltzmann constant where σ = 5.67 × 10^{-8} W/m^2 K^4.

Real surfaces deviate from the ideal behaviour of blackbodies. Specifically, they emit radiation less efficiently than a blackbody, i.e. their spectral emissive power will be lower than the curves plotted in Figure 5.1. They may also emit preferentially at some wavelengths but less so at others, meaning that their curves may be of a different shape than those shown in Figure 5.1. Finally, unlike the diffuse emission illustrated in Figure 5.2 they may emit more in some directions than others.

However, BPS tools do not consider these spectral and directional complexities. Rather, they treat surfaces as *grey* emitters. With this, the surface is considered to emit a fixed fraction of blackbody radiation for all directions and for all wavelengths. The emissive power of a grey surface (E_g) can therefore be determined through a scalar multiple of the emission from a blackbody at the same temperature:

$$E_g = \epsilon_{lw} \cdot E_b$$
$$= \epsilon_{lw} \cdot \sigma \cdot T^4 \tag{5.2}$$

where ϵ_{lw} is the total longwave hemispherical emissivity (−), which has a value in the range of $0 < \epsilon_{lw} < 1$.

Figure 5.3 contrasts the emission from a grey surface to that of a blackbody. As can be seen, the spectral emissive power from the grey surface at each wavelength is equal to that of the blackbody multiplied by ϵ_{lw}.

ϵ_{lw} is a property of a surface. The $_{lw}$ subscript indicates that it is applicable for emission within the longwave spectrum. As explained above, any object within the temperature range of −40 °C to +60 °C emits more than 99 % of its radiation within the longwave spectrum. As all building surfaces operate within this temperature range, this means that BPS tools can treat the emissivity as temperature-invariant.

5.3 RADIATIVE PROPERTIES

Kirchhoff's law of thermal radiation formalized a long-standing experimental observation that good emitters were also good

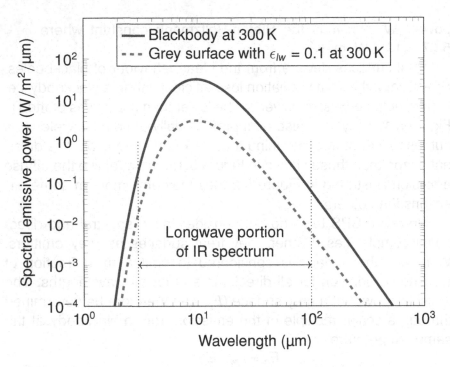

Figure 5.3: Grey-surface versus blackbody spectral emissive power (log-log scale)

absorbers. Consider the perfectly insulated cavity at temperature T_{cavity} depicted in Figure 5.4. The inside of the cavity can be treated as a grey surface. Therefore, the cavity emits radiation according to Equation 5.2. Some of this radiation will be reflected by the walls of the cavity, as illustrated in the figure. A portion will be absorbed, but because the cavity is perfectly insulated this absorbed energy will be re-emitted according to Equation 5.2. The combination of emission, re-emission, and reflection creates a blackbody radiation field within the cavity. As such, the cavity behaves as a blackbody emitting radiation according to Equation 5.1.

Now consider the small object a within the cavity. It is irradiated by the cavity while at the same time it emits radiation according to Equation 5.2. When the system formed by the object and the cavity is allowed to achieve thermal equilibrium, a first law energy balance reveals that object a's emission must equal its absorption of the

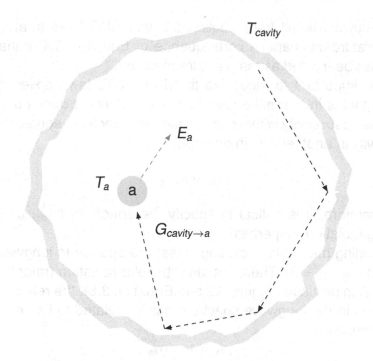

Figure 5.4: Kirchhoff's law

cavity's blackbody irradiation:

$$\cancel{A_a} \cdot E_a = \cancel{A_a} \cdot \alpha_{lw,a} \cdot G_{cavity \to a} \qquad (5.3)$$

Object a's emission (the left side of Equation 5.3) can be represented with Equation 5.2. By recognizing that irradiation of object a is due to blackbody emission at the enclosure's temperature, the right side of Equation 5.3 can be expressed using Equation 5.1. This leads to:

$$\epsilon_{lw,a} \cdot \sigma \cdot T_a^4 = \alpha_{lw,a} \cdot G_{cavity \to a}$$
$$= \alpha_{lw,a} \cdot \sigma \cdot T_{cavity}^4 \qquad (5.4)$$

Since $T_a = T_{cavity}$ under thermal equilibrium conditions, from Equation 5.4 it can be seen that:

$$\epsilon_{lw,a} = \alpha_{lw,a} \qquad (5.5)$$

Therefore, if the emissivity of an object in the longwave spectrum is known, then its absoptivity in the longwave spectrum is also known. Recall from Section 5.2 that BPS tools treat ϵ_{lw} as temperature-invariant. A consequence of Equation 5.4 is that α_{lw} can also be treated as temperature-invariant.

It is important to recognize that Kirchhoff's law applies for an object that is in thermal equilibrium with a cavity. It does not mean that the absorptivity within one wavelength spectrum equals the absorptivity in another. So, in general:

$$\alpha_{lw} \neq \alpha_{solar} \tag{5.6}$$

Therefore, it is critical to specify the spectrum ($solar$, lw) when stating radiation properties.

Building materials (including glass) are opaque to longwave radiation (i.e. $\tau_{lw} = 0$). Therefore, as with solar radiation (refer to Section 3.2, in particular Figure 3.2 and Equation 3.8), the reflectivity of radiation in the longwave spectrum can be related to its emissivity or absorptivity:

$$\rho_{lw} = 1 - \alpha_{lw}$$
$$= 1 - \epsilon_{lw} \tag{5.7}$$

This means that specifying either ϵ_{lw} or α_{lw} is sufficient for defining how a surface will emit, absorb, and reflect longwave radiation given the above assumptions. Some BPS tools require ϵ_{lw} as an input, while others α_{lw}.

5.4 VIEW FACTORS

From the above we see that all surfaces emit, absorb, and reflect radiation. In order to calculate the exchange of radiation between surfaces we need to know how much of the radiation leaving one surface (either by emission or reflection) will reach another surface.

Figure 5.5 illustrates two surfaces that are emitting radiation. The fraction of the radiation emitted or reflected by surface 1 that is incident upon surface 2 is called the *view factor*[i] from surface 1 to surface 2, and is denoted by $f_{1 \to 2}$. Likewise, the view factor $f_{2 \to 1}$ represents the fraction of radiation emitted or reflected by surface

---

[i]The terms *configuration factor* and *shape factor* are also commonly used.

2 that is incident upon surface 1. These view factors are a characteristic of the geometrical relationship between the two surfaces.

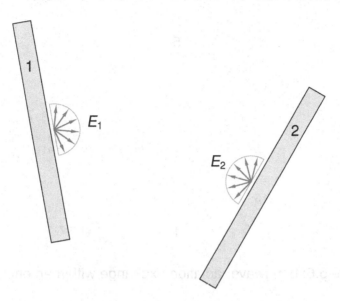

Figure 5.5: Radiation exchange between two surfaces

There is an important view factor relation known as *reciprocity* (refer to any basic heat transfer textbook), which relates the view factors between these (and any other) surfaces:

$$A_1 \cdot f_{1 \to 2} = A_2 \cdot f_{2 \to 1} \tag{5.8}$$

Another important view factor relation is the *summation rule* for the surfaces forming an enclosure:

$$\sum_s f_{1 \to s} = 1 \tag{5.9}$$

where s is the number of surfaces forming the enclosure.

5.5 ENCLOSURE THEORY

The longwave radiation exchange between a zone's internal surfaces can be treated with a special case of radiation theory known as the *diffuse grey enclosure*. This method demands a number of assumptions, which in general are reasonable approximations for radiation exchange between a zone's internal surfaces.

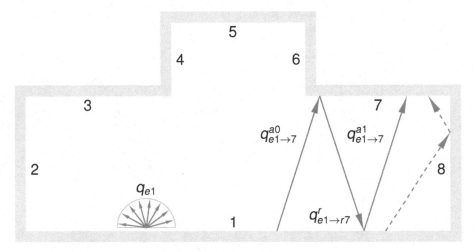

Figure 5.6: Longwave radiation exchange within an enclosure

Figure 5.6 illustrates the plan view of an enclosure formed by 10 surfaces (surfaces 9 and 10, not shown, are the floor and ceiling). The longwave radiation exchange between these 10 surfaces can be resolved with diffuse grey enclosure theory subject to the following assumptions:

- The zone is completely enclosed.
- All surfaces are opaque to longwave radiation.
- The medium within the cavity (air and any occupants, furniture, etc. contained within the zone) does not participate in the radiation exchange.
- Each surface is at a uniform temperature (e.g. all of surface 2 is at T_2).
- All surfaces emit diffusely according to Equation 5.2.
- All surfaces reflect diffusely.
- Each surface is exposed to uniform irradiance (e.g. all of surface 8 receives the same irradiance).
- Each surface obeys Equation 5.5.

5.6 CALCULATING EXCHANGE BETWEEN SURFACES

We will focus on surfaces 1 and 7 of Figure 5.6 to illustrate how enclosure theory is used to calculate the radiation exchange between each pair of surfaces.

We begin by considering the radiation emitted by surface 1 that is absorbed by surface 7 as a result of interactions between only these two surfaces. Since surface 1 is grey, its emitted energy (q_{e1}) is given by Equation 5.2:

$$q_{e1} = A_1 \cdot E_1 = \epsilon_{lw,1}\sigma A_1 T_1^4 \qquad (5.10)$$

q_{e1} will be incident upon all the surfaces that it can view, which is all the other surfaces in Figure 5.6. The portion that will be incident upon surface 7 is given by the view factor $f_{1\to7}$. The $\alpha_{lw,7}$ portion of this will be absorbed by surface 7:

$$
\begin{aligned}
q_{e1\to7}^{a0} &= q_{e1} \cdot f_{1\to7} \cdot \alpha_{lw,7} \\
&= \left[\epsilon_{lw,1}\sigma A_1 T_1^4\right] \cdot f_{1\to7} \cdot \epsilon_{lw,7} \qquad (5.11) \\
&= \left[(\epsilon_{lw,1})(\epsilon_{lw,7})\right] \cdot \left[\sigma A_1 T_1^4\right] \cdot f_{1\to7}
\end{aligned}
$$

since $\epsilon_{lw,7} = \alpha_{lw,7}$ by Equation 5.5.

The ^{a0} superscript denotes the portion of surface 1's emission that was directly absorbed by surface 7 without inter-surface reflection. This is illustrated by the leftmost ⟶ in Figure 5.6.

The $\rho_{lw,7}$ portion of the emission from surface 1 that is incident upon surface 7 will be reflected:

$$
\begin{aligned}
q_{e1\to7}^{r} &= q_{e1} \cdot f_{1\to7} \cdot \rho_{lw,7} \\
&= \left[\epsilon_{lw,1}\sigma A_1 T_1^4\right] \cdot f_{1\to7} \cdot (1 - \epsilon_{lw,7}) \qquad (5.12) \\
&= \left[(\epsilon_{lw,1})(1 - \epsilon_{lw,7})\right] \cdot \left[\sigma A_1 T_1^4\right] \cdot f_{1\to7}
\end{aligned}
$$

since $\rho_{lw,7} = 1 - \epsilon_{lw,7}$ by Equation 5.7. $q_{e1\to7}^{r}$ is the reflected radiation (middle ⟶ in Figure 5.6).

A portion of this reflected radiation will be directed towards surface 1 ($f_{7\to1}$), which will reflect some of it ($\rho_{lw,1} = 1 - \epsilon_{lw,1}$) back towards surface 7 ($f_{1\to7}$). Part of this ($\alpha_{lw,7} = \epsilon_{lw,7}$) will then be absorbed by surface 7:

$$
\begin{aligned}
q_{e1\to7}^{a1} &= q_{e1\to7}^{r} \cdot f_{7\to1} \cdot \rho_{lw,1} \cdot f_{1\to7} \cdot \alpha_{lw,7} \\
&= \left[(\epsilon_{lw,1})(1-\epsilon_{lw,7})\right] \cdot \left[\sigma A_1 T_1^4\right] \cdot f_{1\to7} \cdot f_{7\to1} \cdot (1-\epsilon_{lw,1}) \cdot f_{1\to7} \cdot \epsilon_{lw,7} \\
&= \left[(\epsilon_{lw,1})(\epsilon_{lw,7})(1-\epsilon_{lw,1})(1-\epsilon_{lw,7})\right] \cdot \left[\sigma A_1 T_1^4\right] \cdot f_{1\to7}^2 \cdot f_{7\to1}
\end{aligned}
$$

(5.13)

where the ^{a1} superscript denotes that the emission by surface 1 was absorbed by surface 7 after one set of reflections between these two surfaces (rightmost ⟶ in Figure 5.6).

By extending this analysis to consider infinite interactions between these two surfaces, an expression can be derived to represent the radiation emitted by surface 1 that is absorbed by surface 7 due to any number of interactions between only these two surfaces:

$$
\begin{aligned}
q_{e1\to7}^{a0+a1\cdots} &= q_{e1\to7}^{a0} + q_{e1\to7}^{a1} + q_{e1\to7}^{a2} + \cdots \infty \\
&= \left[(\epsilon_{lw,1})(\epsilon_{lw,2})\right] \cdot \left[\sigma A_1 T_1^4\right] \cdot f_{1\to7} \\
&\quad + \left[(\epsilon_{lw,1})(\epsilon_{lw,7})(1-\epsilon_{lw,1})(1-\epsilon_{lw,7})\right] \cdot \left[\sigma A_1 T_1^4\right] \cdot f_{1\to7}^2 \cdot f_{7\to1} \\
&\quad + \left[(\epsilon_{lw,1})(\epsilon_{lw,7})(1-\epsilon_{lw,1})^2(1-\epsilon_{lw,7})^2\right] \cdot \left[\sigma A_1 T_1^4\right] \cdot f_{1\to7}^3 \cdot f_{7\to1}^2 \\
&\quad + \cdots \infty
\end{aligned}
$$

(5.14)

Equation 5.14 can be represented more compactly as a geometric series:

$$
q_{e1\to7}^{a0+a1\cdots} = \frac{\left[(\epsilon_{lw,1})(\epsilon_{lw,7})\right] \cdot \left[\sigma A_1 T_1^4\right] \cdot f_{1\to7}}{1 - (1-\epsilon_{lw,1})(1-\epsilon_{lw,7}) \cdot f_{1\to7} \cdot f_{7\to1}}
$$

(5.15)

So far we have considered radiation emitted by surface 1 that is absorbed by surface 7 as a result of interactions between only these two surfaces. But some of the radiation emitted by surface 1 can also be absorbed by surface 7 as a result of interactions with other surfaces. Referring to Figure 5.6, emission from surface 1 could, for example, irradiate surface 8, be reflected by surface 8 towards surface 7, and then absorbed by surface 7 (shown by - - -➤ in Figure 5.6).

Using an analysis similar to that presented above for two-surface interactions, it can be shown that the radiation emitted by surface 1 that is absorbed by surface 7 due to infinite three-surface interactions via surface 8 can be given as:

$$q_{e1 \to 8 \to 7}^{a0+a1\cdots} = \frac{\left[(\epsilon_{lw,1})(\epsilon_{lw,7})(1-\epsilon_{lw,8})\right] \cdot \left[\sigma A_1 T_1^4\right] \cdot f_{1 \to 8} \cdot f_{8 \to 7}}{1 - (1-\epsilon_{lw,1})(1-\epsilon_{lw,7})(1-\epsilon_{lw,8}) \cdot f_{1 \to 8} \cdot f_{8 \to 7} \cdot f_{7 \to 1}}$$

(5.16)

An expression like Equation 5.16 can be developed for the seven other possible three-surface interactions for emission from surface 1 that is absorbed by surface 7: $1 \to 2 \to 7 \ldots 1 \to 10 \to 7$.

The total amount of emission from surface 1 that is absorbed by surface 7 can now be determined by summing Equations 5.15, 5.16, and similar expressions to consider any number of surface interactions:

$$q_{e1 \Rightarrow 7} = q_{e1 \to 7}^{a0+a1\cdots} + q_{e1 \to 2 \to 7}^{a0+a1\cdots} + q_{e1 \to 3 \to 7}^{a0+a1\cdots} + \cdots$$

$$= \frac{\left[(\epsilon_{lw,1})(\epsilon_{lw,7})\right] \cdot \left[\sigma A_1 T_1^4\right] \cdot f_{1 \to 7}}{1 - (1-\epsilon_{lw,1})(1-\epsilon_{lw,7}) \cdot f_{1 \to 7} \cdot f_{7 \to 1}}$$

$$+ \sum_{\substack{i=2 \\ i \neq 7}}^{i=10} \left[\frac{\left[(\epsilon_{lw,1})(\epsilon_{lw,7})(1-\epsilon_{lw,i})\right] \cdot \left[\sigma A_1 T_1^4\right] \cdot f_{1 \to i} \cdot f_{i \to 7}}{1 - (1-\epsilon_{lw,1})(1-\epsilon_{lw,7})(1-\epsilon_{lw,i}) \cdot f_{1 \to i} \cdot f_{i \to 7} \cdot f_{7 \to 1}} \right]$$

$$+ \cdots$$

(5.17)

where $q_{e1 \Rightarrow 7}$ represents the total amount of emission from surface 1 that is absorbed by surface 7 due to any number of sets of reflections and due to all possible combinations of surface reflections.

Although Equation 5.17 fully describes the emission from surface 1 that is absorbed by surface 7, it alone is insufficient for calculating the *exchange* of radiation between these surfaces because emission from surface 7 is also absorbed by surface 1. A similar expression can be written for it:

$$q_{e7\Rightarrow1} = q_{e7\to1}^{a0+a1\cdots} + q_{e7\to2\to1}^{a0+a1\cdots} + q_{e7\to3\to1}^{a0+a1\cdots} + \cdots$$

$$= \frac{\left[(\epsilon_{lw,1})(\epsilon_{lw,7})\right] \cdot \left[\sigma A_7 T_7^4\right] \cdot f_{7\to1}}{1 - (1 - \epsilon_{lw,1})(1 - \epsilon_{lw,7}) \cdot f_{1\to7} \cdot f_{7\to1}}$$

$$+ \sum_{\substack{i=2 \\ i\neq7}}^{i=10} \left[\frac{\left[(\epsilon_{lw,1})(\epsilon_{lw,7})(1 - \epsilon_{lw,i})\right] \cdot \left[\sigma A_7 T_7^4\right] \cdot f_{7\to i} \cdot f_{i\to1}}{1 - (1 - \epsilon_{lw,1})(1 - \epsilon_{lw,7})(1 - \epsilon_{lw,i}) \cdot f_{7\to i} \cdot f_{i\to1} \cdot f_{1\to7}}\right]$$

$$+ \cdots$$

(5.18)

Recall from Section 5.1 that we are trying to calculate the *net* rate of radiation exchange from one surface to another: $q_{lw,s\to i}$. Let's consider that we are forming the energy balance for surface 1 ($i = 1$) and that we are calculating the net radiation exchange from surface 7 ($s = 7$) to surface 1. This can be determined by subtracting Equation 5.17 from Equation 5.18:

$$q_{lw,7\to1} = q_{e7\Rightarrow1} - q_{e1\Rightarrow7}$$

$$= \frac{\left[(\epsilon_{lw,1})(\epsilon_{lw,7})\right] \cdot \left[\sigma A_7 T_7^4 \cdot f_{7\to1} - \sigma A_1 T_1^4 \cdot f_{1\to7}\right]}{1 - (1 - \epsilon_{lw,1})(1 - \epsilon_{lw,7}) \cdot f_{1\to7} \cdot f_{7\to1}}$$

$$+ \sum_{\substack{i=2 \\ i\neq7}}^{i=10} \left[\frac{\left[(\epsilon_{lw,1})(\epsilon_{lw,7})(1 - \epsilon_{lw,i})\right] \cdot \left[\sigma A_7 T_7^4\right] \cdot f_{7\to i} \cdot f_{i\to1}}{1 - (1 - \epsilon_{lw,1})(1 - \epsilon_{lw,7})(1 - \epsilon_{lw,i}) \cdot f_{7\to i} \cdot f_{i\to1} \cdot f_{1\to7}}\right]$$

$$- \sum_{\substack{i=2 \\ i\neq7}}^{i=10} \left[\frac{\left[(\epsilon_{lw,1})(\epsilon_{lw,7})(1 - \epsilon_{lw,i})\right] \cdot \left[\sigma A_1 T_1^4\right] \cdot f_{1\to i} \cdot f_{i\to7}}{1 - (1 - \epsilon_{lw,1})(1 - \epsilon_{lw,7})(1 - \epsilon_{lw,i}) \cdot f_{1\to i} \cdot f_{i\to7} \cdot f_{7\to1}}\right]$$

$$+ \cdots$$

(5.19)

With Equation 5.19 we can now calculate the net radiation heat transfer from surface 7 to surface 1. (The equation can be safely truncated as the magnitude of the terms diminishes with increased surface interactions.) But this is only one of the longwave radiation terms in surface 1's energy balance. The internal-surface energy balance of Equation 2.14 includes a summation of such terms because we need to consider the net radiation from all of the other surfaces within the zone. Therefore, to evaluate the energy balance for surface 1 for the case depicted in Figure 5.6 we need to evaluate a total of nine expressions like Equation 5.19. The same holds for forming the energy balances for the zone's other nine surfaces, leading to a total of 90 such equations ($n \cdot [n - 1]$, where n is the number of surfaces within the zone).

Examine the variables involved in Equation 5.19. The surface areas will be determined by the geometrical input provided by the user. The longwave emissivities of each surface will be supplied by the user, either explicitly or implicitly if they select construction materials from tool default databases. You will explore the impact of $\epsilon_{lw,i}$ during this chapter's simulation exercises.

In addition to the unknown surface temperatures, Equation 5.19 also includes view factors between surfaces.

5.7   DETERMINING VIEW FACTORS

The set of 90 expressions of Equation 5.19 required for our case depicted in Figure 5.6 involves 90 view factors. Even by taking advantage of the reciprocity relation (Equation 5.8) this still means that we need to determine 45 unique view factors.

Because view factors depend only upon geometry, they are time-invariant and need only be determined once prior to a timestep simulation. A wide spectrum of methods—some of which you will learn about during this chapter's required reading—are employed by BPS tools to determine these values.

Any basic heat transfer textbook—and certainly more advanced texts on radiation—include tables, figures, and formulas to determine view factors for particular situations. Many of these have been determined through analytical solutions or numerical approximation. However, few of these configurations—coaxial parallel disks, three-sided enclosures, etc.—are appropriate for buildings.

The determination of view factors for more generic geometries, such as the floor plan illustrated in Figure 5.6, invariably relies on some sort of numerical method. Many algorithms employing numerical quadrature have been developed to calculate view factors for radiation exchange between building surfaces (Walton, 2002; Francisco et al., 2014; Kramer et al., 2015). These include various approaches based on area integration or contour integration, which can, in principle, accurately determine view factors for any arbitrary geometry. Efforts have also been expended to minimize the computational burden of these techniques.

Ray-tracing[ii] offers another accurate and general approach. With this, each surface is subdivided into small elemental areas and a unit hemisphere is established above the centre point of each elemental area. These hemispheres are subdivided into patches representing equal solid angles and each solid angle is projected until it intersects with another surface. This process is repeated for each patch of each hemisphere/elemental area to develop a numerical approximation of the view factor between the surface and each other surface bounding the zone. A numerical implementation of this ray-tracing approach, which is based upon a graphical method utilized before the advent of digital computers, is well described in Clarke (2001, Chapter 7).

Unfortunately very few BPS tools support such detailed approaches. Indeed, this is not even a possibility for tools that represent geometry abstractly, wherein users input surface areas and zone volumes rather than calculating these quantities from explicit geometrical input (e.g. vertex coordinates).

Some BPS tools that require explicit geometrical input employ analytical methods to calculate view factors from the user-defined geometry, although these are limited to cases where no surface obstructs another (so-called *convex zones*). These methods cannot resolve situations like Figure 5.6, because, for example, surface 4 obstructs surface 6's view of surface 2.

Some BPS tools allow users to prescribe view factors between surfaces, perhaps after they have been calculated by an external tool. However, such facilities are rarely employed as dozens or even hundreds of view factors per zone may be required for even simple geometrical configurations.

Some simplistic approximations based upon area-weighting

[ii] Also known as the *unit-sphere method*.

approaches are also in use, such as:

$$f_{i \to j} = \frac{A_j}{\sum_k A_k - A_i} \qquad (5.20)$$

where the $\sum_k$ summation considers all the surfaces in the zone.

The advantage of this method is that explicit geometrical information is not required from the user since it only considers the area of surfaces and not their exact shape. Although computationally efficient, this method provides an exact answer only for cubes. It can lead to erroneous view factors for some common situations such as co-planar surfaces. For example, Equation 5.20 gives a non-zero view factor between a window and the wall which contains it.

Some tools refine this approximation by setting view factors to co-planar surfaces (such as the window and its containing wall) to zero and by limiting the summation in the denominator of Equation 5.20 to "seen" (i.e. non-co-planar surfaces). Although an improvement, this method also suffers from the limitation that it only produces exact results for a cube.

Users should be aware of the default and optional methods available in their BPS tool for estimating view factors and should be aware of the consequences of any simplifications. In some cases more accurate view factors can be determined with minimal effort simply by invoking optional facilities. During this chapter's simulation exercises you will explore which method your chosen BPS tool uses by default and which optional methods it supports.

5.8 LINEARIZATION AND SIMPLIFICATIONS

Section 2.8 discussed how BPS tools solve the large set of coupled equations that describe indoor mass and energy transfers. The two approaches applied by most contemporary BPS tools use direct or iterative procedures to solve a linearized set of equations.

Section 5.6 detailed how the net longwave radiation between a pair of surfaces can be calculated using Equation 5.19. However, since this equation is non-linear (it includes temperatures to the fourth power), it must be linearized to fit within the solution process.

Linearization is accomplished by expressing the heat transfer between the surfaces using a longwave radiation coefficient (h_{lw}):

$$q_{lw,7 \to 1} = A_7 \cdot h_{lw,7 \to 1} \cdot (T_7 - T_1) \tag{5.21}$$

This coefficient is determined by equating Equations 5.19 and 5.21:

$$
\begin{aligned}
h_{lw,7 \to 1} = \frac{1}{A_7\,(T_7 - T_1)} \cdot \Bigg\{ & \frac{\big[(\epsilon_{lw,1})(\epsilon_{lw,7})\big] \cdot \big[\sigma A_7 T_7^4 \cdot f_{7 \to 1} - \sigma A_1 T_1^4 \cdot f_{1 \to 7}\big]}{1 - (1-\epsilon_{lw,1})(1-\epsilon_{lw,7}) \cdot f_{1 \to 7} \cdot f_{7 \to 1}} \\
& + \sum_{\substack{i=2 \\ i \neq 7}}^{i=10} \left[\frac{\big[(\epsilon_{lw,1})(\epsilon_{lw,7})(1-\epsilon_{lw,i})\big] \cdot \big[\sigma A_7 T_7^4\big] \cdot f_{7 \to i} \cdot f_{i \to 1}}{1 - (1-\epsilon_{lw,1})(1-\epsilon_{lw,7})(1-\epsilon_{lw,i}) \cdot f_{7 \to i} \cdot f_{i \to 1} \cdot f_{1 \to 7}} \right] \\
& - \sum_{\substack{i=2 \\ i \neq 7}}^{i=10} \left[\frac{\big[(\epsilon_{lw,1})(\epsilon_{lw,7})(1-\epsilon_{lw,i})\big] \cdot \big[\sigma A_1 T_1^4\big] \cdot f_{1 \to i} \cdot f_{i \to 7}}{1 - (1-\epsilon_{lw,1})(1-\epsilon_{lw,7})(1-\epsilon_{lw,i}) \cdot f_{1 \to i} \cdot f_{i \to 7} \cdot f_{7 \to 1}} \right] \\
& + \cdots \Bigg\}
\end{aligned}
$$

$$\tag{5.22}$$

Therefore, the overall solution process incorporates Equation 5.21, and Equation 5.22 is used to calculate its required coefficients. As explained in Section 2.8, some BPS tools will recalculate the h_{lw} coefficients using updated temperatures as they iterate to a converged solution within the timestep, while other tools will avoid this recalculation by evaluating the coefficients using the temperatures solved for the previous timestep. As explained in Section 2.8, BPS tools employing the latter approach may be more sensitive to the user's choice of simulation timestep.

It is important to note that not all BPS tools treat longwave radiation exchange in the rigorous fashion detailed in this chapter. To reduce the computation burden, some tools introduce a fictitious temperature[iii] and use it to calculate the approximate net radiation exchange between an internal surface and all other surfaces. This replaces the $\sum_s q_{lw,s \to i}$ summation in the internal surface energy

---

[iii]Often called the *star* temperature.

balance with a single term that accounts for the net radiation exchange into surface i from all other internal surfaces. However, this approximation deviates from the exact result when the zone is not cubic; or when not all internal surfaces are at the same temperature; or when not all surfaces have the same $h_{conv,i}$ convection coefficients.

Other tools further simplify the problem by eliminating the $\sum_s q_{lw,s \to i}$ altogether from the internal-surface energy balance and by increasing the convection coefficient $h_{conv,i}$ to account for the combined effects of convection and longwave radiation heat transfer. The implicit assumption that the temperature of the other internal surfaces in the zone (T_s) are the same as the zone air temperature (T_z) may introduce significant errors in some situations.

5.9 REQUIRED READING

Reading 5–A

Francisco *et al.* (2014) presents an algorithm for calculating view factors for complex building geometries based upon Stokes' theorem.

Read pages 203–205 of this article and find answers to the following:

1. What are the classes of methods they list that have been used to calculate view factors for generic geometries?
2. Why is the double integral (Equation 7 in the article) necessary for calculating view factors using area integration?
3. What are the advantages of using Stokes' theorem to convert these area integrals to line integrals?
4. Why does their method rely on subdividing surfaces into elemental areas and then determining view factors between these elemental areas?

5.10 SOURCES FOR FURTHER LEARNING

- Walton (2002) describes and contrasts the performance of several area integration and line integration methods for calculating view factors. He also describes how these have been implemented in the freely available View3D program that can be used to calculate view factors between internal building surfaces. This tool is shipped with EnergyPlus as an "auxiliary" program.

- Chapter 7 of Clarke (2001) details ESP-r's ray-tracing method for calculating view factors.
- Kramer *et al.* (2015) describe how they have borrowed methods from the computer graphics domain to create a computationally efficient algorithm to calculate area-to-area view factors.

5.11 SIMULATION EXERCISES

Revert your BPS tool inputs to represent once again the *Base Case* described in Section 1.9, including any refinements or corrections you made following the previous exercises. Perform an annual simulation to produce a fresh set of *Base Case* results for use in the following exercises.

Exercise 5–A

Consult the technical documentation for your BPS tool to determine its options for calculating longwave radiation exchange between internal surfaces. Does it explicitly calculate $q_{lw,s \to i}$ for each pair of surfaces? If not, what approach does it use? (Refer to Section 5.8.) Are there options to invoke more detailed approaches?

If for your *Base Case* your tool was configured to employ a simplified method rather than calculating $q_{lw,s \to i}$ for each pair of surfaces, then invoke its most detailed approach and perform another annual simulation using the same timestep.

What impact does this change have upon the annual space heating load? And upon the annual space cooling load?

Exercise 5–B

Longwave emissivity values were specified for the materials forming the internal surfaces of all the opaque envelope constructions in the *Base Case*. Now reduce the concrete floor's ϵ_{lw} from 0.9 to 0.1 and perform another annual simulation using the same timestep. This is the value of ϵ_{lw} that might be expected for a polished metal. If your BPS tool does not allow you to modify a surface's ϵ_{lw} then consult its technical documentation to determine why.

What impact does this change have upon the annual space heating load? And upon the annual space cooling load? Are these results in line with your expectations?

What are possible sources of information that can be used to determine ϵ_{lw} values for materials? Describe a situation in which accurately establishing a material's longwave emissivity may have an important impact upon simulation predictions.

Exercise 5–C

Undo the change made in Exercise 5–B.

What is the default method employed by your BPS tool for determining radiation view factors between internal surfaces? (Refer to Section 5.7.) How did you configure your *Base Case* to determine view factors? If you did not use the default method provided by your BPS tool, then configure your tool to now employ its default method.

Now perform another simulation in which your BPS tool applies its most detailed method for determining radiation view factors between internal surfaces. Describe the algorithm employed by your BPS tool and contrast the view factors determined in the two simulations.

What impact does this change have upon the annual space heating load? And upon the annual space cooling load? Describe a situation in which accurately establishing these view factors may have an important impact upon simulation predictions.

5.12 CLOSING REMARKS

This chapter explained that BPS tools which explicitly calculate longwave radiation exchange between internal surfaces treat the surfaces as isothermal and as diffuse grey emitters and reflectors. With this most rigorous treatment, the user is responsible for prescribing the emissivity (or absorptivity) of all internal surfaces in the longwave portion of the infrared spectrum (or rely on tool default values). This chapter also explained that many BPS tools use simplified methods to treat longwave radiation exchange in order to reduce the computational burden. You developed an understanding of which methods your chosen BPS tool supports through the simulation exercises.

Through the required reading and the theory presented in this chapter, you learnt about the types of methods that are available for determining view factors between surfaces. You became aware of which methods are available in your chosen BPS tool through the simulation exercises, and realized the actions required to invoke them.

CHAPTER 6

Internal heat and moisture sources

Tʜɪs chapter describes how sources of heat and moisture within the building are treated within the indoor environment energy and mass balances, and discusses the significance of establishing appropriate input parameters.

Chapter learning objectives

1. Understand that both the magnitude and the schedule of internal heat gains and moisture sources are typically prescribed by the user.
2. Become aware of commonly used sources of data for establishing these inputs.
3. Realize the sensitivity of simulation predictions to user choices and the range of uncertainty over these inputs.
4. Understand the possibilities and complexities of modelling occupant behaviour as an alternative to the user prescribing internal heat and moisture sources.

Table 6.1: Heat and moisture sources within the interior of the building

Heat sources	Occupant metabolic activity Lighting Computers, printers, scanners, etc. Entertainment equipment, chargers, etc. Clothes washers and dryers (machine) Dishwashers, refrigerators, etc.
Moisture sources	Occupant respiration and perspiration Bathing and showering Cleaning, dishwashing, cooking Unvented gas cookers Clothes drying (hang dry) Indoor plants

6.1 HEAT AND MOISTURE SOURCE TERMS IN MASS AND ENERGY BALANCES

Chapter 2 described how mass and energy balances are formed for the building interior. It was seen that the zone moisture mass balance (Equation 2.6) included a summation term representing the sources of water vapour within the zone ($\dot{m}_{v,source}$). Summation terms representing internal heat sources appeared in both the zone and internal surface energy balances. The convective portion of these heat sources ($q_{conv,source \to z}$) appeared in the zone energy balance (Equation 2.13) while the radiant portion ($q_{lw,source \to i}$) appeared in the internal surface energy balance (Equation 2.14).

This chapter discusses how the $\dot{m}_{v,source}$, $q_{conv,source \to z}$, and $q_{lw,source \to i}$ terms are determined.

6.2 SOURCES OF HEAT AND MOISTURE IN BUILDINGS

Some of the many possible sources of heat and moisture within the interior of the building are listed in Table 6.1.

The terms appearing in the balance equations represent summations of all possible contributions. In this way, $\sum \dot{m}_{v,source}$ is the sum of all the moisture sources listed in Table 6.1 (and perhaps others). $\sum q_{conv,source \to z}$ represents the sum of the convective components of the metabolic activity of the occupants, lighting, computers, appliances, etc., while $\sum q_{lw,source \to i}$ represents the sum of the radiative components of these gains. It is left to the user (or tool default methods) to determine the convective/radiative split of each heat source and to determine how to distribute the radiative sources to the energy balance of each internal surface.

Since the mass and energy balances are formed and solved each timestep of a simulation, it is necessary for the user to define which gains are present in each zone, the magnitude of these gains, and how these gains vary in time.

6.3   MAGNITUDE OF GAINS

Various sources are available for estimating the magnitude of heat and moisture gains. A popular one is ASHRAE's Handbook of Fundamentals (ASHRAE, 2017). Chapter 18 of this handbook, which is focused on nonresidential cooling and heating load calculations, provides numerous tables summarizing data from several sources to estimate gains from occupants, lights, computers, and some appliances. Chapter 6 of CIBSE Guide A (CIBSE, 2015) is another well-documented source commonly consulted by BPS users that provides similar tabulated data on internal gains.

Data from these ASHRAE and CIBSE tables can be used to estimate moisture and heat gains from occupants. They estimate, for example, the sensible heat gain (the sum of $q_{conv,source \to z}$ and $q_{lw,source \to i}$) of an adult male seated in an office at a dry-bulb temperature of 24 °C to be 75–80 W. An adult female in a comparable situation is estimated to give 65–70 W of heat gain. These values can increase or decrease by up to 20 % depending upon the room temperature.

According to the ASHRAE table, the apportioning of these sensible heat gains to convective ($q_{conv,source \to z}$) and radiative ($q_{lw,source \to i}$) components depends upon the air velocity over the person. At low air velocities, 40 % of the heat gains can be considered convective and 60% radiative; at higher air velocities the split is taken to be 73% convective and 27% radiative.

Data are provided for many other activities, such as moderately active office work, sedentary work, light bench work, heavy work in a factory, etc. The challenge for the BPS tool user often relates to answering questions such as: How many people are in the thermal zone? What are those people doing? Are they men, women, or children? What split between convective and radiative components should be assumed?

The ASHRAE and CIBSE tables also provide "latent heat" gains of occupants. For example, the latent heat gain of occupants conducting office work is in the range of 40 to 70 W. This is defined to equal the rate of energy released during a constant-temperature (at T_z) phase transition of water from vapour to liquid:

$$q_{latent,source \to z} = \dot{m}_{v,source} \cdot h_{fg}|_{T_z} \qquad (6.1)$$

where $q_{latent,source \to z}$ is the latent heat gain (W) and $h_{fg}|_{T_z}$ is the enthalpy of vaporization of water (J/kg) at the zone temperature.

It is common for BPS tools to require users to prescribe moisture sources in terms of latent gains. In this case, Equation 6.1 is used to calculate the value of $\dot{m}_{v,source}$ that is required by the mass balance based upon the user's prescribed value for $q_{latent,source \to z}$. Again, the challenge for the BPS tool user is to answer questions such as: What is the level of activity of the occupants? How are they clothed? How many occupants are in the zone? Are they men, women, or children?

The ASHRAE and CIBSE (and other) documents provide similar data on gains from lights, computers, and appliances, but again the BPS tool user must answer many questions in order to select appropriate data. What type of lights are used: fluorescent, incandescent, compact fluorescent, LED? Are luminaires installed within the thermal zone, or mounted on ceilings? If the latter, what fraction of the heat output is added to the zone under consideration, and what fraction to the ceiling plenum (if treated as a separate thermal zone)? Are desktop or laptop computers used? What is their nominal power draw? How many lights and computers are installed in the zone? Etc.

You will explore the sensitivity of BPS predictions to some of these decisions during this chapter's simulation exercises.

6.4 TEMPORAL VARIATION OF GAINS

As the mass and energy balances are formed and solved each timestep of the simulation, $\dot{m}_{v,source}$, $q_{conv,source \to z}$, and $q_{lw,source \to i}$ are actually functions of time. Consequently, the BPS tool needs to know when occupants are present and when lights and other equipment are functioning so that the magnitudes of the gains treated in the previous section can be scaled.

Most commonly the user prescribes schedules (profiles) of activities for this purpose. For example, the user might indicate that occupants begin arriving at an office building at 6h30, that full occupancy is achieved by 9h00, that half the occupants leave the building over the lunch hour, and that there is no occupancy overnight. Such schedules, which may vary by days of the week or seasonally, are time-series of scalar multipliers that operate on the user-prescribed gain magnitudes. For example, if 350 W is prescribed as the magnitude of the convective heat gain from occupants in a zone and the prescribed schedule indicates that 70 % of these occupants are present at 10h30 on a given day, then the simulation will use a value of 245 W for the occupant portion of $q_{conv,source \to z}$ for that timestep of the simulation.

There are several possibilities for establishing these schedules. With existing buildings it might be feasible to measure occupancy over short periods of time to develop bespoke schedules for a particular building. However, when dealing with buildings in the design or pre-occupancy phase, it is common for users to establish schedules based upon their own experience or to rely on published guidelines and recommendations.

One frequently used source of schedules is the US Department of Energy's Commercial Reference Building (CRB) database (Deru et al., 2011). This provides representative schedules of internal gains based on monitoring experience for various commercial building types (office, hotel, warehouse, etc.). Another common source of schedules is COMNET, an initiative of the New Buildings Institute to standardize BPS approaches for codes and standards. Some codes and standards (e.g. NECB, 2017; NCM, 2015) also provide representative schedules for various building types. Although these are meant for demonstrating compliance with regulations, their use is widespread for other types of BPS analyses as these schedules are viewed by many BPS users as being representative.

It is important for the user to recognize that whatever source they use for establishing schedules, their decision will have an impact on the values of $\dot{m}_{v,source}$, $q_{conv,source \to z}$, and $q_{lw,source \to i}$ that are used in the balance equations. Additionally, the BPS tool's reported consumption of energy for servicing lighting, computers, and other plug loads will be directly calculated from these user-selected inputs.

Figure 6.1 plots lighting schedules for office buildings from three of the aforementioned sources. Although these three profiles share a good deal of similarity, they are not identical and therefore will lead to different simulation results. This figure also plots the median schedule for one particular office building that was monitored by Bennet and O'Brien (2017). From this, it is clear that a BPS analysis of this office building using either the NECB, CRB, or COMNET profiles would lead to substantial overprediction of $q_{conv,source \to z}$ and $q_{lw,source \to i}$ during most of the daytime.

You will explore the impact of schedules of internal gains during this chapter's simulation exercises.

6.5   OCCUPANT BEHAVIOUR

Most of the sources of moisture and heat gains within buildings are influenced by occupant behaviour. Occupant arrival and departure times and metabolic activity, the occupant switching of lights, and operating computers and appliances all impact the $\dot{m}_{v,source}$, $q_{conv,source \to z}$, and $q_{lw,source \to i}$ terms that are required by the moisture and energy balances.

Many authors have shown that the use of prescribed schedules to characterize all of these complex behaviours can lead to significant prediction errors in particular cases. Many BPS tools provide facilities that allow the user to represent occupant behaviour on a

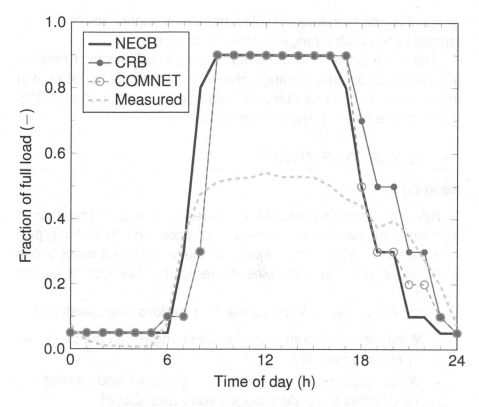

Figure 6.1: Sample lighting schedules

simplified level. These can be used, for example, to switch off lights when daylight levels exceed a certain threshold.

There has been considerable research activity within the BPS field to develop more detailed models to predict occupant behaviour as an alternative to users prescribing schedules or simple binary actions. Most of these so-called *occupant behaviour* models are developed from observational studies and involve some sort of stochastic modelling approach. It is common to use a predictor variable (e.g. indoor air temperature, illuminance) to estimate the probability of some occupant behaviour (e.g. opening a window, closing a blind) based upon a randomly selected number. Given the stochastic nature of these models, BPS simulations are not repeatable and therefore multiple simulations might need to be performed to establish a range of possible outcomes.

This is an active area of research, as you will discover through this chapter's required readings. However, to date the application of these methods has been far less common and most users of BPS rely on some sort of prescribed schedules.

6.6   REQUIRED READING

Reading 6–A

Gunay *et al.* (2016) summarize a number of models that have been developed to consider the behaviour of occupants in building performance simulation. The majority of these occupant models use explanatory (predictor) variables to predict the likelihood of a state change.

Read this article and find answers to the following questions:

1. What are some examples of state changes that are predicted by existing models?
2. Which explanatory variables are typically used predict the likelihood of the state changes you listed above?
3. What is a Markovian occupant model?
4. Why are some models limited to specific simulation timesteps?
5. How does the choice of occupancy model influence simulation predictions?
6. Explain why occupant behaviour models are rarely used by BPS practitioners.

Reading 6–B

O'Brien *et al.* (2017) provide a guest editorial for a special issue of the *Journal of Building Performance Simulation* on the fundamentals of occupant behaviour research.

Read this article and find answers to the following questions:

1. What are the three main stages of occupant research?
2. What are the key unresolved research questions related to occupant behaviour?

6.7 SOURCES FOR FURTHER LEARNING

- Chapter 5 of Deru *et al.* (2011) describes numerous sources of data for establishing inputs for internal heat and moisture sources. It also thoroughly documents the assumptions and decisions behind the magnitudes and schedules for internal gains given in the US Department of Energy's Commercial Reference Building (CRB) database.
- Mahdavi and Tahmasebi (2019) discuss how occupants can influence building performance and describe the various options for modelling their behaviour.

6.8 SIMULATION EXERCISES

Revert your BPS tool inputs to represent once again the *Base Case* described in Section 1.9, including any refinements or corrections you made following the previous exercises. Perform an annual simulation to produce a fresh set of *Base Case* results for use in the following exercises.

Exercise 6–A

Time-invariant (constant) internal gains that are 100 % sensible and 100 % convective were prescribed for the *Base Case*.

Alter the internal heat gains such that they are 100 % sensible and 100 % radiative and perform another annual simulation. What impact does this have upon the annual space heating load? And upon the annual space cooling load? Explain this result by referring back to the energy balances developed in Chapter 2, referring in particular to Equations 2.13 and 2.14.

Discuss the impact of accurately determining the split of convective and radiative output from an internal heat source. What are possible sources of information that can be used to determine these parameters?

Exercise 6–B

Revert to your *Base Case*.

Now, rather than employing constant internal gains as in the

Base Case, configure your simulation to represent the case where the internal gains are caused by the presence of:

- three occupants
- three desktop computers with flat-panel monitors
- 200 W of desk-mounted fluorescent lighting

Assume that the occupants are present and the computers and lights used only between 9h00 and 12h00 and between 13h00 and 17h00 Mondays through Fridays. Use an appropriate source (see Section 6.3) to establish the magnitude of the heat and moisture gains caused by the occupants, computers, and lights, including the convective and radiative split.

Contrast the results of this simulation and those of the *Base Case*. What impact has this change had upon the annual space heating load? And upon the annual space cooling load? Estimate the range of uncertainty in the magnitude of the $\dot{m}_{v,source}$, $q_{conv,source \rightarrow z}$, and $q_{lw,source \rightarrow i}$ terms that you chose?

Exercise 6–C

Continue with the assumptions you took in Exercise 6–B, but now perform another simulation in which the work day commences one hour earlier (i.e. at 8h00), but still ends at 17h00.

What impact does this have upon the annual space heating load? And upon the annual space cooling load? Comment on the impact that uncertainty over occupant arrival has upon simulation predictions.

Exercise 6–D

Revert to the inputs used for Exercise 6–B. Now increase your estimates of the magnitude of the occupant heat and moisture gains by 10 %. Also increase your estimate of the magnitude of computer heat gains by 10 %. Perform another simulation.

What impact does this have upon the annual space heating load? And upon the annual space cooling load? Comment on what impact the variability between occupants and the estimates of heat gains from office equipment may have on the uncertainty of simulation predictions.

6.9   CLOSING REMARKS

This chapter explained that the user must prescribe both the magnitude and the schedule of internal heat and moisture sources. Decisions are also required to divide the heat sources between the energy balances representing the zone and those representing internal surfaces. Some of the sources of data that can be used to establish these important inputs were mentioned, and you became familiar with at least one of them by conducting the simulation exercises. These exercises also helped you become aware of the impact of these user choices. And finally, through the required readings you learnt that considerable research is underway to establish models of occupant behaviour to reduce some of the uncertainty in treating internal heat and moisture sources.

CHAPTER 7

Internal airflow

THIS chapter discusses methods that are used in the BPS field for treating inter-zone airflow. It also touches upon approaches that can be used to predict airflow patterns and the distribution of air temperatures and contaminant concentrations within zones.

Chapter learning objectives

1. Realize the common options for treating airflow between zones.
2. Understand the conceptual basis of network airflow models, zonal models, and computational fluid dynamics (CFD) models.
3. Learn how to configure your chosen BPS tool to consider inter-zone airflow.
4. Realize the impact of inter-zone airflow upon simulation predictions.

7.1 TRANSFER OF AIR BETWEEN ZONES

The term $\dot{m}_{a,t-in}$ appears in the zone dry-air mass balance (Equation 2.3), the zone moisture mass balance (Equation 2.6), and the zone energy balance (Equation 2.13). This term represents air flowing into the zone under consideration from other zones within the building. This inter-zone airflow is known as *transfer air*. $\dot{m}_{a,t-in}$ appears within a summation in each of these balances because the zone under consideration may be receiving transfer air from numerous other zones.

Air may be transferred from one zone to another intentionally or not. As implied by Equation 2.3, air transfers between zones are closely related to air infiltration and airflow forced by HVAC systems (ventilation).

7.2 OPTIONS FOR TREATING TRANSFER AIR

There are five common approaches used in BPS for establishing $\dot{m}_{a,t-in}$ values:

1. It is very common for users to ignore air transfers between zones, essentially setting all $\dot{m}_{a,t-in}$ values to zero. This is the default treatment with most BPS tools, one that the user can override in most cases.

2. Users prescribe time-invariant values for each possible flow path.

3. Users prescribe scheduled values, which perhaps vary by time of day or by season.

4. Some BPS tools provide facilities that can be used in conjunction with one of the above options to make airflow rates vary in consequence to a simulation parameter, such as an indoor air temperature.

5. Predicting airflow rates for all of the possible flow paths with a ventilation model.

You will explore which of these approaches are supported by your chosen BPS tool during this chapter's simulation exercises.

7.3 VENTILATION MODELS

CFD is a modelling approach that is widely applied in predicting building ventilation performance. With this, a building zone is subdivided into small volumes (many thousands or even millions). The equations that govern the conservation of mass, momentum, and energy are written and approximated for each volume using discretization methods. This yields a large set of equations whose solution leads to a prediction of airflow patterns and temperature and contaminant distributions within a zone.

Some BPS tools include integrated facilities that couple the calculation of building energy flows (using the approaches we have studied in previous chapters) with CFD modelling of one or multiple zones of the building. However, due to CFD's high requirements for computational resources, and the significant expertise and time required to correctly configure turbulence options, boundary conditions, etc., such facilities are rarely employed in simulating the performance of whole buildings.

You will learn more about the application of CFD to predict building performance during this chapter's required reading.

Zonal models offer another option for predicting ventilation. Like CFD models, they subdivide a zone into a number of smaller volumes. However, they are not as computationally demanding as CFD because they apply a much coarser subdivision. Although many zonal models have been developed based upon empirical and analytical relations, these approaches are rarely used in simulating the performance of whole buildings or in predicting transfer airflow rates between zones.

The only modelling approach that enjoys widespread use in the BPS field for predicting $\dot{m}_{a,t-in}$ values are the so-called *multi-zone network airflow models*. Typically each thermal zone of the building is represented by a single airflow *node* (although subdivision is possible). The nodes—each of which represents a well-mixed volume of uniform conditions—are connected together by the user into a network that defines possible flow paths.

The user also prescribes *components* for each path to describe its flow characteristics. Typically these component equations relate the flow along a path (such as transfer air) to the pressure difference between nodes, for example:

$$\dot{m}_{a,t\text{-}in} = \mathcal{F}\,(P_{t\text{-}in} - P_z) \tag{7.1}$$

P_z is the pressure at the node representing the zone under consideration and $P_{t\text{-}in}$ is the pressure at a node of the zone that is the source of the transfer air. Various forms of the function $\mathcal{F}$ are in use, and these must be imposed by the user.

An instance of Equation 7.1 is established for each of the possible flow paths, and appropriate boundary conditions are imposed. The solution of this set of equations then leads to the transfer airflow rates required by the zone mass and energy balances.

In most cases when nodal airflow modelling is applied, the network considers both airflow through the building envelope (infiltration and natural ventilation) and transfer air between zones. Consequently, these models will be described in greater detail in Chapter 15 in the context of air infiltration and natural ventilation modelling. That chapter will explain how the nodal pressures are calculated, the options for the $\mathcal{F}$ functions, and the treatment of boundary conditions.

7.4 REQUIRED READING

Reading 7–A

Chen (2009) provides an overview of methods for predicting ventilation performance.

Read this article in its entirety and find answers to the following questions:

- With the well-mixed assumption, a thermal zone is considered to have uniform conditions (temperature and chemical composition). This is an appropriate assumption for what types of building spaces according to the author? And when might this assumption be inappropriate?
- This article describes seven categories of methods for predicting ventilation performance, three of which are simulation methods. What are these three simulation methods?

- Which of the three simulation methods is most commonly used for predicting the ventilation performance of entire buildings (as opposed to individual rooms)?
- What are the two popular network airflow models mentioned in the article?
- Why does the author believe that zonal models have limited applicability?
- Most of the applications of CFD for predicting ventilation performance fall into three categories. What are these three categories? Why is CFD often employed for these types of analyses?

7.5 SOURCES FOR FURTHER LEARNING

- Megri and Haghighat (2007) describe the basic principles of zonal models, their development, and their application.
- Srebric (2019) describes methods for predicting airflow, air temperature, and contaminant concentrations in buildings. She provides, in particular, a good introduction to CFD theory, including the treatment of turbulence and boundary conditions.
- Cook *et al.* (2003) discuss the application and validation of CFD for predicting the temperature stratification within a room exposed to combined wind and buoyancy forces.

7.6 SIMULATION EXERCISES

Revert your BPS tool inputs to represent once again the *Base Case* described in Section 1.9, including any refinements or corrections you made following the previous exercises. Perform an annual simulation to produce a fresh set of *Base Case* results for use in the following exercises.

Now revert your BPS tool inputs to the two-zone representation of Exercise 2–E. Include the same refinements or corrections you made to the *Base Case*. Perform an annual simulation to produce a fresh set of Exercise 2–E results for use in the following exercises.

Exercise 7–A

Recall that there was no airflow between the two thermal zones in Exercise 2–E. This is the first method listed in Section 7.2. Now impose a constant transfer airflow rate of 100 L/s between the zones (the second method) and perform another simulation using the same timestep.

Create a temperature-versus-time graph for March 6. Plot the zone air temperatures from the three simulations on this graph. Do the results from the simulation with the imposed transfer airflow more closely resemble those from the *Base Case* or those from Exercise 2–E?

What impact does the addition of transfer airflow have upon the annual space heating load? And upon the annual space cooling load?

What are possible sources of information that can be used to establish plausible transfer airflow rates? In Exercise 2–E you were asked in what situations the additional effort and risk of input errors associated with subdividing the building into thermal zones would be justified. Think about this question anew in light of these results.

Exercise 7–B

Explore the other facilities offered by your chosen BPS tool for treating transfer air. Can you schedule these airflow rates, or control them based upon simulation parameters, such as the zone air temperature? Does your tool support a network airflow model, a zonal model, or an integrated CFD model?

Invoke one or more of these optional facilities and perform some additional simulations. Contrast the results with those from Exercise 7–A.

7.7 CLOSING REMARKS

This chapter explained that although inter-zone airflows are commonly ignored by BPS users, there are modelling options for predicting these values. Through the required reading and the material provided in this chapter, you learnt about the conceptual basis of network, zonal, and CFD airflow models. You also learnt about the applicability of each. And you developed an appreciation for the impact of inter-zone airflow rates upon simulation predictions during this chapter's simulation exercises.

This completes our treatment of heat and mass transfer processes occurring within the interior of the building. Our attention now turns to the exterior environment.

III

Exterior environment

CHAPTER 8

Energy balances at external surfaces & weather

THIS part of the book focuses on heat transfer processes relevant to the exterior environment. An energy balance at external surfaces is developed in this chapter, and then the important topic of weather data is treated.

Chapter learning objectives

1. Become cognizant of all the terms appearing in the energy balances representing external surfaces.
2. Appreciate the influence of exterior environmental conditions on all the mass and energy balances seen thus far.
3. Understand where weather files come from and their limitations.
4. Appreciate how typical meteorological year weather files are composed and when they should be used.
5. Realize the sensitivity of simulation predictions to weather data.

8.1 ENERGY BALANCE AT EXTERNAL SURFACES

Some of the individual heat and mass transfer processes occurring between a building and its surrounding environment were discussed in Chapter 1 (see Section 1.1). These types of heat transfer and mass transfer processes were then illustrated schematically in Figure 2.1. Chapter 2 described approaches for forming mass and energy balances for the building interior. Our attention now turns to the energy balances that are formed for each external surface of the building.

Consider the external surfaces illustrated in Figure 2.1. Each above-grade (above-ground) surface will experience convection heat transfer with the outdoor air and may absorb solar radiation, depending upon the sun position. Each of these surfaces will also exchange longwave radiation with objects in the external environment that it views: the ground surface, surrounding buildings and structures, atmospheric gases and clouds, and outer space. And heat will be transferred from the mass of the materials comprising the envelope assemblies to these external surfaces (or the reverse). Surfaces in contact with the ground (below-grade) will also exchange heat via conduction with the surrounding soil.

All of these heat transfer processes are illustrated schematically for the single external surface depicted in Figure 8.1, where an infinitesimally thin control volume is drawn around the surface. Drawing upon the methods detailed in Chapter 2, a first law energy balance can be formed for this control volume:

$$q_{solar \to e} + q_{cond,m \to e} = q_{conv,e \to oa} + \sum_{env} q_{lw,e \to env} + q_{e \to grd} \qquad (8.1)$$

The subscript e represents the external surface under consideration. All terms in this equation are expressed in W. $q_{solar \to e}$ is the rate at which incident solar radiation is absorbed by the surface. $q_{cond,m \to e}$ is the rate of heat transfer from the mass of the envelope construction into surface e. $q_{conv,e \to oa}$ is the rate of convection heat transfer from surface e to the outdoor air. $q_{lw,e \to env}$ is the net exchange of longwave radiation from surface e to an object in the exterior environment env. This term appears in a summation be-

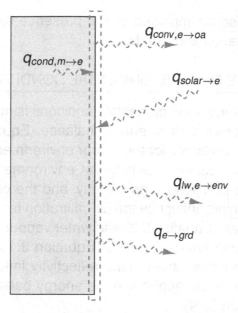

Figure 8.1: Energy balance at external surface e (control volume indicated by - - - - -)

cause surface e will exchange radiation with a number of entities, as explained above.

The $q_{solar \rightarrow e}$, $q_{conv,e \rightarrow oa}$, and $q_{lw,e \rightarrow env}$ terms are only pertinent for above-grade surfaces. For below-grade surfaces, $q_{e \rightarrow grd}$ represents the heat transfer from surface e to the surrounding soil.

An energy balance in the form of Equation 8.1 is formed for each external surface. These are combined with the mass and energy balances describing the indoor environment and the building envelope and then collectively solved using the methods elaborated in Section 2.8 for each timestep of the simulation.

Although the methods used to represent each of the terms in Equation 8.1 will only be treated in later chapters, the importance of exterior environmental conditions on this energy balance is obvious. The position of the sun, conditions in the lower and upper atmosphere, and ground reflectivity (influenced by snow cover) will affect $q_{solar \rightarrow e}$, while $q_{conv,e \rightarrow oa}$ will depend upon the outdoor air temperature as well as wind velocity adjacent to the surface. Soil temperatures will influence the $\sum_{env} q_{lw,e \rightarrow env}$ and $q_{e \rightarrow grd}$ terms, while the former will also be impacted by the presence and temperature of atmospheric gases and clouds.

8.2 INFLUENCE OF ENVIRONMENTAL CONDITIONS

The impact of exterior environmental conditions is not limited to the terms of the external surface energy balance (Equation 8.1). The mass and energy balances for the indoor environment that we have already seen also require knowledge of environmental conditions. The atmospheric pressure, wind velocity, and the outdoor air temperature and humidity influence the air infiltration terms of the mass balances on dry air (Equation 2.3) and water vapour (Equation 2.6), as well as the zone energy balance (Equation 2.13). And the sun position, sky conditions, and ground reflectivity influence the term for solar absorption that appears in the energy balance for internal surfaces (Equation 2.14).

Table 8.1 summarizes the influence that environmental conditions have on the mass and energy balances we have seen thus far. Ideally, all of these influencing environmental parameters would be measured at the building site and at a time interval equal to or smaller than the simulation timestep. In this way the BPS tool could, for example, predict air infiltration (the subject of Chapter 15) with minimal uncertainty by using the wind velocity adjacent to each building surface at each timestep. And longwave radiation exchange at external surfaces could be accurately computed using the measured temperature of participating atmospheric gases.

However, this ideal is never realized in practice because some of the environmental conditions listed in Table 8.1 are not commonly measured. Moreover, rarely are environmental measurements taken at the building site—an impossibility in the case of buildings at the design stage. Instead BPS users typically specify a *weather file* whose contents directly or indirectly specify the necessary boundary conditions.

8.3 WEATHER FILES AND THEIR LIMITATIONS

Weather is defined to be the state of the atmosphere at a given location. This includes the movement of air; its pressure, temperature,

Table 8.1: Influence of exterior environmental conditions on mass and energy balances

Process	Term	Equation	Influencing conditions
Solar absorption	$q_{solar \rightarrow e}$	8.1	Sun position
	$q_{solar \rightarrow i}$	2.14	Atmospheric scattering
			Ground reflectivity
			Trees and other shading objects
Air infiltration	$\dot{m}_{a,inf}$	2.3	Wind velocity
	$(\omega \cdot \dot{m}_a)_{inf}$	2.6	Air temperature
	$(\dot{m}_a \cdot c_P')_{inf} \cdot (T_{inf} - T_z)$	2.13	Humidity
			Atmospheric pressure
Surface convection	$q_{conv,e \rightarrow oa}$	8.1	Wind velocity
			Air temperature
Longwave radiation	$q_{lw,e \rightarrow env}$	8.1	Cloud cover
			Moisture as function of altitude
			Temperature of ground surfaces
			Temperature of surrounding objects
Ground heat transfer	$q_{e \rightarrow grd}$	8.1	Soil temperature

and humidity; precipitation; and the type and extent of cloud cover in the troposphere and beyond.

Motivated by a desire to improve the safety of marine navigation, governments in some parts of the world formalized the observation and forecasting of weather in the 19th century. The instrumentation and techniques for observing weather evolved over time, and important new uses for these data emerged. Thanks to these weather observations and forecasts, farmers could better plan crop plantings and harvests, the safety of marine and air navigation improved, and people could plan their daily activities with more certainty.

The previous section made clear the importance of the knowledge of environmental conditions to BPS models. Sadly, these needs have never been given a great deal of consideration by the weather observation community. As stated by Crawley and Barnaby (2019), BPS has always had to "make do" with whatever data has been gathered for other purposes.

For starters, most weather data available to BPS users have only been logged at one-hour intervals, despite the fact that some weather parameters—in particular wind velocity and cloud cover—can fluctuate at much higher frequencies. Usually the prevailing conditions measured at the top of the hour (e.g. 13h00) are logged.

Figure 8.2, which plots wind speed measurements that were taken at a building site, illustrates the drawback of hourly data. The wind speed measurements were taken at 1-minute intervals (indicated by ○ symbols). However, only the values measured at the top of each hour (indicated by ■ symbols) would be logged in an hourly weather file and therefore available to a BPS tool. When BPS tools are operated with sub-hourly timesteps (as you may have been doing in the simulation exercises), the hourly values from the weather file are usually interpolated to estimate the weather parameters for the prevailing timestep. A linear interpolation of the hourly data is also shown in the figure (———). This is how most BPS tools would represent wind speed as a boundary condition. The poor correspondence between the hourly interpolation and the 1-minute data that is clearly seen in Figure 8.2 means that wind speed—and other highly fluctuating weather conditions—can be inaccurately represented when hourly weather files are employed.

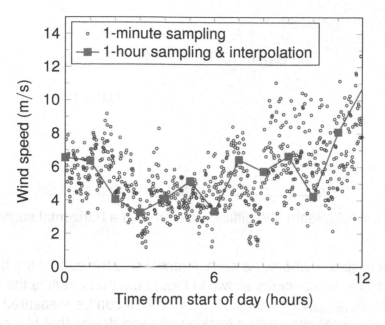

Figure 8.2: Wind speed measured at a building site: comparison of interpolation of hourly data to measurements at 1-minute intervals

The treatment of solar radiation is another issue. As indicated in Table 8.1, knowledge of atmospheric scattering is required to calculate the solar absorption terms for surface energy balances. This is accomplished using two measurements of solar radiation, which can be combined to determine the beam and diffuse irradiance to horizontal surfaces. Recall from Section 3.3 that diffuse radiation has been scattered by the earth's atmosphere while beam radiation has not. The beam and diffuse irradiance components are illustrated schematically in Figure 8.3.

Weather files either contain direct normal irradiance (DNI) and diffuse horizontal irradiance (DHI), or global horizontal irradiance (GHI) and DHI. DNI can be measured with a pyrheliometer mounted on a tracking device that keeps the instrument aimed at the sun at all times. In this way, the device measures only the beam irradiance of Figure 8.3 to a surface whose normal is kept aligned with the beam: $G_{solar\_beam \rightarrow \perp}$. GHI can be measured with a horizontally mounted pyranometer that has an unobstructed view of

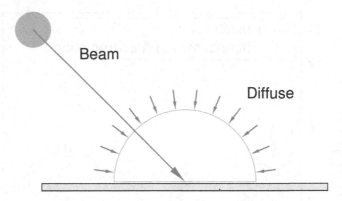

Figure 8.3: Beam and diffuse irradiance to a horizontal surface

the hemisphere above. As such, it represents the sum of the beam and diffuse components shown in Figure 8.3 that irradiate the horizontal: $G_{solar\_beam\rightarrow\_}$ + $G_{solar\_diff\rightarrow\_}$. And DHI can be measured with a pyranometer fitted with a tracking shading device that blocks the beam component from striking the sensor. As such, it measures $G_{solar\_diff\rightarrow\_}$.

Unlike most other weather parameters, which are logged at the top of the hour, the solar radiation values contained in standard weather files represent integrated quantities over the previous hour, e.g. the value logged at 15h00 is the integrated quantity from 14h00 to 15h00. Although this does not allow BPS tools to consider the impact of short-term effects, such as fast-moving clouds, at least the boundary conditions consider all the solar energy received over the hour.

That said, unfortunately the vast majority of weather files comes from weather stations that lack pyrheliometers and pyranometers. As a consequence, in most cases the DNI, GHI, and DHI data in weather files are not directly measured, but rather are estimated by models that predict these quantities from observed cloud cover data. Although such modelling methods can provide reasonably accurate estimates on a daily integrated basis, the errors on an hourly basis can be significant. In fact, the root-mean-square errors of the modelling of hourly GHI using such approaches has been observed to be in the order of 30 % (Crawley and Barnaby, 2019). Newer models that operate on satellite observations rather than station-based cloud observations offer the potential to improve the accuracy of solar radiation data in weather files (Huang, 2019), so this is an area that will hopefully be improved in the future.

Another important issue is the absence of measurements of some of the required boundary conditions listed in Table 8.1. Although some weather file formats include fields for ground reflectivity, very few measurements are available in practice so this field is rarely populated. As well, most of the environmental parameters required to calculate longwave radiation from external surfaces are absent from standard weather files, and rarely do the files contain data on soil temperatures in the vicinity of a building.

It is important for the user to understand the inherent limitations of weather data and the uncertainty this contributes towards BPS analyses. The task for the user usually comes down to making an appropriate choice amongst the available weather files.

8.4   WHICH WEATHER FILE?

Which weather file to use depends greatly upon the objectives of the BPS analysis. For starters, you should choose a weather file from a station that is located in geographical proximity to the building site, or at least from one that has similar meteorological conditions. Keep in mind that most weather stations are located at airports where conditions may differ from nearby urban or suburban settings due to urban heat island effects, which often result in warmer temperatures in cities compared to surrounding rural areas. Additionally, wind velocities are strongly influenced by local topography and the placement of surrounding buildings and structures.

For some types of analyses we want to supply the BPS tool with boundary conditions that replicate as closely as possible conditions that were actually observed for a given period of time. In such cases the most appropriate choice would be an *actual meteorological year* (AMY) weather file. Such AMY files are available for thousands of locations—particularly in North America and Europe—in some cases from the middle of the 20th century to the present time. Users should investigate the sources of the data contained in these files, and be aware that they may contain some interpolated or modelled data (in addition to solar radiation) that are used to replace erroneous or missing values since a file that contains records for each hour of the year is normally required.

Year-to-year weather conditions can vary significantly. Some years are warmer, some cooler, some sunnier, some cloudier. Indeed, no two years share identical weather patterns. For some analyses we might want to assess the performance of a building and its energy systems under the hottest and sunniest conditions, or under the coldest and cloudiest. In this case it might be necessary to choose amongst the available AMY files, or perhaps to conduct simulations with many different AMY files to examine performance under varying conditions.

In many cases we want to assess performance under "typical" conditions. This is such a common need that significant efforts have been expended to generate files that represent the *typical meteorological year* (TMY)[i]. Although these are commonly called weather files, they are in fact files that characterize *climate* rather than describing observed weather. Climate can be thought of as a long-term (typically 30 years) average of weather.

TMY files are structured to contain typical data for each hour of an artificial year. As such, they can be used to estimate typical performance using single-year simulations. They are composite files that contain measured (or modelled) weather data drawn from many years. It is common to start with several decades of AMY data and to compose the TMY file one month at a time. Using a statistical process, the most typical January from the set of AMY files is selected, then the most typical February, and so on. The statistical process used to select the most typical months considers cumulative probability distributions of parameters such as minimum, maximum, and mean quantities of temperature, humidity, wind speed, and solar radiation.

Most of the weather files available to BPS users have been developed using the TMY procedure. But it is important to realize that there is no universal scheme for composing TMY files. The parameters considered and the weights placed upon their importance

---

[i]Sometimes called *test reference year* (TRY).

vary, so a given location may have multiple TMY files, each composed from AMY data drawn from different time periods.

The selection procedure used to compose TMY files tends to eliminate months with more extreme weather. As such, despite their ubiquity, users should not rely on simulations conducted with TMY files to assess design robustness, to estimate peak loads, or to predict performance under extreme conditions. That said, single-year simulations conducted with TMY files can predict long-term energy consumption with reasonable accuracy (Crawley and Barnaby, 2019).

In addition to AMY and TMY files, there have also been efforts to create or adjust weather files to account for urban heat island effects and future climate change, which you will explore in greater depth during this chapter's required readings and during one of the simulation exercises. Some weather files are also composed to assess extreme weather conditions for testing building resiliency using BPS. It is worth noting that these types of weather files are far less common, and much less readily available than TMY and AMY files.

8.5   SOURCES OF WEATHER FILES

Most BPS tools are shipped with a basic set of weather files, but in many cases the user will have to acquire additional data.

The EnergyPlus weather (EPW) file format was introduced more than two decades ago (Crawley et al., 1999) to support the needs of the BPS community and has become an unofficial standard, being supported by more than 25 BPS tools. EPW is a file format, not a source or type of weather file, and as such is used for AMY, TMY, and other file types. Although the file format supports sub-hourly data, the vast majority of EPW files available to BPS users contain hourly records.

In addition to tool-specific websites, users can find EPW formatted weather files from a number of sources. Climate.OneBuilding.Org offers thousands of EPW (and other) formatted weather files. Most are TMYs that have been generated using various methods and base AMY files. Another source is White Box Technologies, which offers TMY files as well as AMY files for the past 20 years for thousands of locations worldwide.

In addition to these two sources, internet searches can reveal countless other sites offering AMY, TMY, and future weather files. Users are cautioned to carefully check the provenance of data and to ensure that files contain full data records for an entire year, because some BPS tools may produce unintended consequences when they encounter missing weather file records.

8.6   REQUIRED READING

Reading 8–A

Bueno *et al.* (2013) describe a model for calculating air temperatures inside urban areas based on measurements taken from nearby weather stations located in open areas (e.g. airports).

Read the introduction to this article and answer the following:

1. At what times of day does the urban heat island effect result in the greatest warming of the air within cities compared to surrounding rural areas?
2. What are four physical factors that contribute to the urban heat island effect?
3. Which building types are most affected by the urban heat island effect?

Reading 8–B

Belcher *et al.* (2005) present a method they call *morphing* to produce weather data for BPS that accounts for future changes to climate.

Read the first three sections of this article and answer the following:

1. What is a global circulation model?
2. Why do the outputs of global circulation models have to be *downscaled* to produce BPS weather files?
3. What are three advantages of the morphing downscaling approach?

4. What is the *shift* morphing technique for adjusting weather file parameters?

5. What is the *stretch* morphing technique for adjusting weather file parameters?

8.7   SOURCES FOR FURTHER LEARNING

- Crawley and Barnaby (2019) provide a detailed discussion of how environmental boundary conditions are treated in BPS. They discuss the characteristics of weather, criteria for selecting weather files, weather data formats, TMYs, methods used to model solar radiation and to fill missing records, and data sources and tools, among other topics.
- Pernigotto *et al.* (2019) propose a method for creating extreme reference year weather files for testing building resiliency using BPS.
- Lauzet *et al.* (2019) review the methods currently available for coupling urban climate models and BPS to better represent local microclimate effects.
- Crawley and Lawrie (2015) argue for a change in the common practice of employing TMY weather data in BPS analyses.

8.8   SIMULATION EXERCISES

Revert your BPS tool inputs to represent once again the *Base Case* described in Section 1.9, including any refinements or corrections you made following the previous exercises. Perform an annual simulation to produce a fresh set of *Base Case* results for use in the following exercises.

Exercise 8–A

Create a temperature-versus-time graph for March 6. Plot the outdoor air temperature on this graph. Superimpose on this graph the temperature of the external surface of the south wall and the temperature of the external surface of the roof. Do the predicted external surface temperatures agree with your expectations?

Based upon this graph, how do you think the magnitude and

direction of the convection heat transfer between the south wall and the outdoor air and between the roof and the outdoor air will vary over the day?

Exercise 8–B

Create a second graph for plotting the rate of heat transfer (W) versus time for the external surface of the roof on March 6. Extract the simulation predictions for the following heat transfer rates and plot these on the graph:

- Absorbed solar radiation, $q_{solar \to e}$
- Convection to the outdoor air, $q_{conv,e \to oa}$
- Net longwave radiation exchange to the external environment, $\sum_{env} q_{lw,e \to env}$
- Heat transfer from the mass of the roof's envelope assembly, $q_{cond,m \to e}$

From this graph examine the magnitude and direction of the convection and longwave radiation terms over the course of the day. How do the $q_{conv,e \to oa}$ results compare with the prediction you made in Exercise 8–A?

Although the methods used to calculate the longwave radiation exchange with the external environment have not yet been explained (this will be treated in Chapter 11), based upon your graph how sensitive do you think the external surface energy balance presented in Section 8.1 is to the accurate calculation on this term? Predict the impact of a 10 % overestimation or underestimation of this term.

Exercise 8–C

You have conducted all the previous simulation exercises using the 2016 version of the Ottawa *Canadian Weather for Energy Calculations* weather file (*CWEC-2016*). This is a TMY file released by the Meteorological Service of Canada that was composed from AMY weather files from 1998 to 2014 using the procedures outlined in Section 8.4.

The previous version of CWEC released by the Meteorological

Service of Canada was composed from older weather data that were gathered from the 1950s to the 1990s (*CWEC-old*). Acquire a copy of the CWEC-old Ottawa weather file for this exercise[ii]. Also, acquire a copy of the AMY weather file for 1998 (*AMY-1998*), the warmest year during the 1990s decade[iii].

Using your BPS tool or a separate utility[iv] determine the average annual outdoor air temperature in the CWEC-2016 weather file as well as the coldest and hottest temperatures over the year. Also, determine the global horizontal irradiance integrated over the year in GJ. Repeat this analysis for the CWEC-old and AMY-1998 weather files.

How do the average, minimum, and maximum outdoor air temperatures vary between these three weather files? How does the annually integrated global horizontal irradiance vary between these three weather files? What impact do you expect these variations to have upon the annual space heating load? And upon the annual space cooling load?

Exercise 8–D

Perform three simulations using the CWEC-2016, CWEC-old, and AMY-1998 weather files. Compare the annual space heating loads and annual space cooling loads from these three simulations. Are these results in agreement with your expectations from Exercise 8–C?

What is your explanation for the differences between the CWEC-2016 and CWEC-old simulations? Think of a situation where it would be appropriate to use AMY weather data rather than TMY weather data.

[ii]This can be downloaded from Climate.OneBuilding.Org or the book's companion website.

[iii]This can be downloaded from the Meteorological Service of Canada or the book's companion website.

[iv]The National Renewable Energy Laboratory's DView and Big Ladder Software's Elements tools are both freely available and handy for such tasks.

Exercise 8–E

You learned about the morphing method to produce weather data to account for future climate change during Reading 8–B. In this exercise you will apply the shift and stretch morphing techniques to the CWEC-2016 weather file to examine the impact of future climate change upon BPS predictions.

Examine one possible future climate scenario by increasing the ambient air temperature by 1 °C for each hour of the year. This can be accomplished by manipulating the CWEC-2016 weather file using the Elements software mentioned in Exercise 8–C. Use Elements' *offset* tool to accomplish this shifting morph. Perform a simulation with this morphed weather file and determine the annual space heating load and the annual space cooling load.

Now examine another possible future climate scenario by reducing the GHI by 2% for each hour of the year. Use Elements' *scale* tool to accomplish this stretch morph. Perform another simulation with this morphed weather file and determine the annual space heating load and annual space cooling load.

Compare these results with those produced in Exercise 8–D. Do the results agree with your expectations? Comment on the uncertainty of future weather trends and how this should be considered in conducting BPS analyses.

8.9   CLOSING REMARKS

This chapter described how energy balances are formed at external building surfaces. Like the mass and energy balances we previously examined for the building interior, these external surface energy balances contain terms that are evaluated by models at each timestep of the simulation.

The remaining chapters of Part III of the book describe most of these models. Chapter 9 explains how the solar radiation absorbed at external surfaces ($q_{solar \to e}$) is calculated. Chapter 10 shows how the $q_{conv,e \to oa}$ term representing convection heat transfer between external surfaces and the outdoor air is evaluated, while the radiation exchange between external surfaces and the exterior environment in the longwave spectrum ($\sum_{env} q_{lw,e \to env}$) is dealt with in

Chapter 11. Finally, Chapter 12 discusses how heat transfer to the surrounding soil ($q_{e \to grd}$) is treated.

All of the models used to calculate these heat transfer terms require boundary conditions that describe the state of the exterior environment. This chapter explained that these exterior environmental conditions are directly or indirectly determined from data contained in a weather file chosen by the user. The above-mentioned chapters will detail how these weather data are used.

Some of the limitations—most importantly, hourly resolution and lack of solar radiation measurements—of the weather files available to BPS users were elaborated in this chapter. The distinctions between AMY, TMY, and other weather file types were explained and guidance was given on selecting appropriate files. Through the chapter's simulation exercises you developed an appreciation for the impact of weather data on predictions.

CHAPTER 9

Solar energy absorption by external surfaces

THIS chapter describes how solar irradiance to external building surfaces is determined based upon the data contained in weather files.

Chapter learning objectives

1. Understand the modelling options available for calculating the solar irradiance of external building surfaces.
2. Understand how the solar radiation data contained in weather files are used in these calculations.
3. Appreciate the inherent uncertainties associated with estimating the contributions of diffuse and ground-reflected solar irradiance.
4. Understand the types of models available for calculating the impact of shading by building elements and external objects.
5. Discover the optional methods for calculating solar irradiance and shading that are available in your chosen BPS tool and their impact.

9.1 MODELLING APPROACH

The external surface energy balance expressed by Equation 8.1 includes the term $q_{solar \rightarrow e}$. This is the rate at which solar radiation is absorbed by the surface under consideration. In a manner similar to the treatment of solar radiation absorption at internal surfaces (Chapter 3), $q_{solar \rightarrow e}$ can be calculated from the incident irradiance:

$$q_{solar \rightarrow e} = \alpha_{solar,e} \cdot A_e \cdot G_{solar \rightarrow e} \tag{9.1}$$

where $G_{solar \rightarrow e}$ is the solar irradiance (W/m^2) incident upon the surface.

As discussed in Section 1.1, the direction and magnitude of this irradiance depends on many factors: scattering by the earth's atmosphere, the geometrical relationship between the building and the sun, shading by surrounding objects, reflection off the ground, etc. It is common for BPS tools to treat the global incident irradiance as the summation of three components:

$$G_{solar \rightarrow e} = G_{solar\_beam \rightarrow e} + G_{solar\_diff \rightarrow e} + G_{solar\_grd \rightarrow e} \tag{9.2}$$

$G_{solar\_beam \rightarrow e}$ is the solar irradiance that has travelled along a direct beam from the sun to the building surface without being scatted by the earth's atmosphere and without being reflected off the ground or other surfaces[i]. This component of irradiance has a single angle of incidence with the building surface, although this angle will, of course, change throughout the day and throughout the year. $G_{solar\_diff \rightarrow e}$ is the solar irradiance that has been scattered by the earth's atmosphere and as such does not have a single angle of incidence. $G_{solar\_grd \rightarrow e}$ is the solar irradiance that is directed to the surface after reflecting off the ground.

Substituting Equation 9.2 into Equation 9.1 yields:

$$q_{solar \rightarrow e} = \alpha_{solar,e} \cdot A_e \cdot \left(G_{solar\_beam \rightarrow e} + G_{solar\_diff \rightarrow e} + G_{solar\_grd \rightarrow e} \right) \tag{9.3}$$

This equation implicitly assumes that the three components of

---

[i]It is sometimes known as *direct* irradiance.

irradiance are uniform over the surface. This will not be strictly correct when, for example, a portion of the surface is shaded by another part of the building or by a surrounding object. However, this assumption is necessary given that the energy balance of Equation 8.1 is formed for the external surface in its entirety.

A_e in Equation 9.3 is the surface area (m^2) which is determined from the geometrical input provided by the user. $\alpha_{solar,e}$ is the total hemispherical absorptivity in the solar spectrum. Values for each surface are supplied by the user. In some tools users may not be aware that they are prescribing the $\alpha_{solar,e}$ values because if they select construction materials from the tool's default databases they may be accepting the $\alpha_{solar,e}$ values chosen by the tool developers. You will explore the impact of $\alpha_{solar,e}$ during one of this chapter's simulation exercises.

The remaining variables that need to be established in order to calculate the required $q_{solar \to e}$ heat transfer rate to each external surface using Equation 9.3 are, therefore, $G_{solar\_beam \to e}$, $G_{solar\_diff \to e}$, and $G_{solar\_grd \to e}$. These quantities will vary from surface to surface, and in time.

The first step in calculating these irradiance quantities is to determine the beam and diffuse irradiance to a horizontal surface based on the data contained in the weather file.

9.2   BEAM AND DIFFUSE IRRADIANCE TO THE HORIZONTAL

Recall from Section 8.3 that weather files specify solar radiation either in terms of DNI ($G_{solar\_beam \to \perp}$) and DHI ($G_{solar\_diff \to \_}$), or in terms of GHI ($G_{solar\_beam \to \_} + G_{solar\_diff \to \_}$) and DHI. These two approaches are alternate ways of specifying equivalent data.

This can be seen by recognizing that DNI is measured normal to the solar beam while GHI and DHI are measured on a horizontal plane (refer to Figure 8.3). In the case of GHI/DHI data, the beam radiation to the horizontal surface can be determined from:

$$G_{solar\_beam \to \_} = GHI - DHI \qquad (9.4)$$

Trigonometry can be used to relate the result of Equation 9.4

to DNI if the angle between the solar beam and the horizontal is known:

$$\sin \eta_{alt} = \frac{G_{solar\_beam\rightarrow\_}}{G_{solar\_beam\rightarrow\perp}} \tag{9.5}$$

η_{alt} is the angle between the solar beam and the horizontal. Known as the *solar altitude*[ii], this angle is illustrated in Figure 9.1 for a horizontal surface placed at a latitude (Φ) in the northern hemisphere.

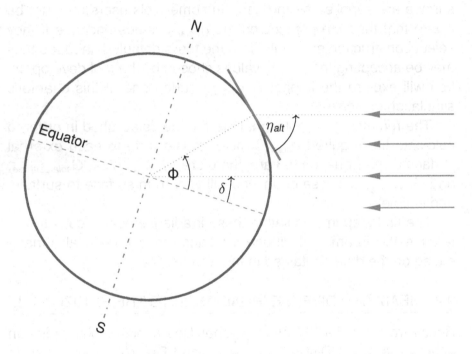

Figure 9.1: Latitude (Φ), solar declination (δ), and solar altitude (η_{alt}) on a day at solar noon

The solar altitude can be determined from three fundamental angles of solar radiation geometry that describe the location of the horizontal surface on earth, the earth's rotation about its axis, and the earth's orbital rotation about the sun: latitude, hour angle, and declination. The latitude is input by the user, either directly or through the selection of a weather file.

[ii]Also referred to as the solar elevation angle, it is the complement of the solar zenith angle.

The earth rotates 15°/h about its axis (one full rotation per day). The *solar hour angle* (Ω) represents the angular displacement between the sun's beam radiation and the local meridian, where for example, Ω is −15° at 11h00 solar time and Ω is 82.5° at 17h30 solar time. BPS tools calculate Ω from the time of day and the site's longitude, usually with adjustments to account for the discrepancy between solar noon and local noon due to perturbations in the earth's rotational speed caused by its orbital path around the sun.

The earth's axis of rotation is tilted at an angle of 23.45° relative to it's orbital plane about the sun. As a result, a direct line drawn from the centre of the sun to the centre of the earth will bisect points in the northern hemisphere during the northern hemisphere's summer, and will bisect points in the southern hemisphere during the southern hemisphere's summer.

During the summer the apparent motion of the sun is such that it rises earlier in the day, reaches a higher point at noon, and sets later. The *solar declination angle* (δ) is defined to be the angle between the sun's beam radiation at solar noon and the equatorial plane. This is illustrated in Figure 9.1. Over the course of the year it varies within the range of $-23.45° < \delta < 23.45°$. It is commonly approximated by BPS tools as a sinusoidal function over the year, although more complex functions that account for the earth's elliptical orbit about the sun are sometimes employed.

The solar altitude can be calculated from these three fundamental angles (Kreith and Kreider, 1978):

$$\eta_{alt} = \sin^{-1}\left[\sin\Phi \cdot \sin\delta + \cos\Phi \cdot \cos\delta \cdot \cos\Omega\right] \qquad (9.6)$$

As δ varies throughout the year and Ω varies throughout the day, Equation 9.6 can be used to calculate the solar altitude angle as a function of time. For instance, Figure 9.2 plots η_{alt} calculated with Equation 9.6 over the course of three days for a latitude of 50°N. As can be seen, compared to November and March, the sun rises earlier in July and attains a much higher solar altitude at noon[iii].

Most BPS tools use the above procedures to calculate the

[iii] SunEarthTools provides a sun position tool that is very helpful in understanding the apparent motion of the sun.

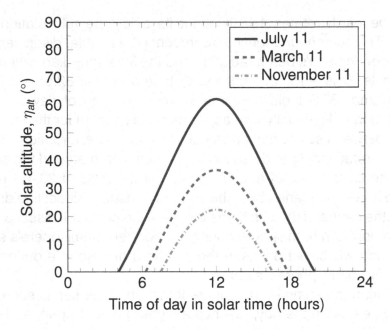

Figure 9.2: Solar altitude angle versus time of day for a latitude of 50 °N

beam and diffuse irradiance to the horizontal. They either calculate $G_{solar\_beam\rightarrow\_}$ using Equation 9.4, or they calculate it using Equations 9.5 and 9.6. In both cases $G_{solar\_diff\rightarrow\_}$ is mapped directly from the weather file's DHI.

The following sections describe the techniques used by BPS tools to determine the three components of irradiance to the external surface that are required by Equation 9.3 from $G_{solar\_beam\rightarrow\_}$ and $G_{solar\_diff\rightarrow\_}$.

9.3 BEAM IRRADIANCE TO A BUILDING SURFACE

It is necessary to determine the orientation of the surface under consideration relative to the solar beam in order to calculate $G_{solar\_beam\rightarrow e}$. The orientation of the surface can be defined using the two angles illustrated in Figure 9.3: the surface slope and azimuth.

The *slope* of the surface (β) is the angle between the plane of

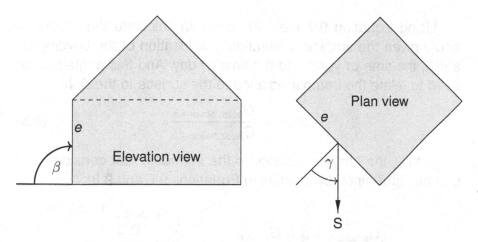

Figure 9.3: Slope and azimuth angles of surface *e*

the surface and the horizontal as seen in an elevation view of the building (left side of Figure 9.3). For example, $\beta = 90°$ for a vertical wall and $\beta = 0°$ for a flat roof.

The surface *azimuth* (γ) is the angle formed between the horizontal projection of the surface normal and the local meridian as seen in a plan view of the building (right side of Figure 9.3). In the northern hemisphere a wall facing due south has $\gamma = 0°$ and a wall facing due east has $\gamma = -90°$.

BPS tools that represent building geometry explicitly will determine β and γ from the geometry data provided by the user (e.g. vertex coordinates). Those that represent geometry abstractly will require the user to provide β and γ as input values.

The angle formed between the surface normal and the solar beam can be determined from these surface orientation angles and the three fundamental angles of solar radiation geometry that were previously defined (Kreith and Kreider, 1978):

$$
\begin{aligned}
\theta = cos^{-1} \Big[& \sin\delta \sin\Phi \cos\beta - \sin\delta \cos\Phi \sin\beta \cos\gamma \\
& + \cos\delta \cos\Phi \cos\beta \cos\Omega + \cos\delta \sin\Phi \sin\beta \cos\gamma \cos\Omega \\
& + \cos\delta \sin\beta \sin\gamma \sin\Omega \Big]
\end{aligned}
$$

$$(9.7)$$

where θ is known as the *incidence* angle.

Using Equation 9.7 the BPS tool can calculate the incidence angle given the surface orientation, the location of the building on earth, the time of year, and the time of day. And this angle can be used to relate the beam irradiance to the surface to the DNI:

$$\cos \theta = \frac{G_{solar\_beam \rightarrow e}}{G_{solar\_beam \rightarrow \perp}} \tag{9.8}$$

Finally, the beam irradiance to the surface under consideration can be determined by combining Equations 9.5 and 9.8:

$$G_{solar\_beam \rightarrow e} = G_{solar\_beam \rightarrow \_} \cdot \frac{\left[\dfrac{G_{solar\_beam \rightarrow e}}{G_{solar\_beam \rightarrow \perp}} \right]}{\left[\dfrac{G_{solar\_beam \rightarrow \_}}{G_{solar\_beam \rightarrow \perp}} \right]} \tag{9.9}$$

$$= G_{solar\_beam \rightarrow \_} \cdot \frac{\cos \theta}{\sin \eta_{alt}}$$

Most BPS tools employ relations similar to Equation 9.6 at each timestep of the simulation to calculate the solar altitude, and relations like Equation 9.7 to determine the incidence angle of the beam radiation to each external surface. As these depend only upon the geometrical relationship between the surface under consideration and the sun, apart from the uncertainly in determining $G_{solar\_beam \rightarrow \_}$ from weather file data (as discussed in Chapter 8) most BPS tools can accurately determine the first irradiance component required by Equation 9.3 using Equation 9.9.

9.4 DIFFUSE IRRADIANCE TO A BUILDING SURFACE

The term $G_{solar\_diff \rightarrow e}$ in Equation 9.3 accounts for the absorption of solar radiation that has been scattered by the earth's atmosphere. As such, it represents the absorption by the surface under consideration of irradiance coming from many directions over the sky dome.

The directional distribution of this diffuse irradiance depends upon the turbidity of the atmosphere, including effects from clouds, aerosols, smog, etc. Duffie and Beckman (2006) present measured data that illustrate the variability of the diffuse solar intensity across the sky dome. On clear days the intensity is greatest in the region around the sun disk. This is caused by forward scattering and is commonly called *circumsolar* diffuse irradiance. Based upon their measurements, Perez *et al.* (1990) observed that this forward scattering is most pronounced under conditions of thin or scattered clouds, or when the atmosphere has a high aerosol content. There also tends to be increased intensity near the horizon under clear sky conditions, an effect caused by Rayleigh scattering and is known as *horizon brightening*.

Figure 9.4 illustrates the circumsolar and horizon brightening effects. These are often thought of as superimposed on top of a sky dome that is irradiating isotropically (equally in all directions).

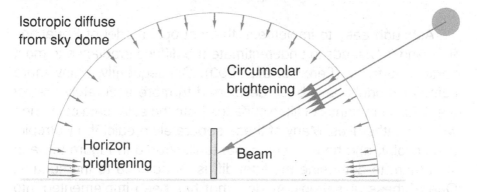

Figure 9.4: Diffuse irradiance from sky dome, circumsolar region, and horizon

Many models have been developed to represent these complex phenomena in order to predict the diffuse solar irradiance to surfaces of arbitrary orientation, that is $G_{solar\_diff \to e}$. The simplest of these, which is used by many BPS tools, is the isotropic model, which essentially ignores circumsolar and horizon brightening. The diffuse irradiance is treated as being of uniform intensity from all directions. As such, the diffuse solar irradiance to the surface under consideration can be easily determined from $G_{solar\_diff \to \_}$ once the view factor (refer to Section 5.4) from the sky dome to the surface has been determined:

$$A_e \cdot G_{solar\_diff \to e} = f_{sky\_dome \to e} \cdot \left(A_{sky\_dome} \cdot G_{solar\_diff \to \_} \right) \quad (9.10)$$

Making use of reciprocity (Equation 5.8), this can be rearranged to express the diffuse irradiance to the surface in terms of the surface's view factor to the sky dome:

$$G_{solar\_diff \to e} = \frac{A_{sky\_dome}}{A_e} \cdot f_{sky\_dome \to e} \cdot G_{solar\_diff \to \_} \qquad (9.11)$$

$$= f_{e \to sky\_dome} \cdot G_{solar\_diff \to \_}$$

The view factor from the surface to the sky dome depends only upon the surface's slope:

$$f_{e \to sky\_dome} = \frac{1 + \cos \beta}{2} \qquad (9.12)$$

Although easy to implement, the isotropic model of Equations 9.11 and 9.12 tends to underestimate the diffuse irradiance to most vertical surfaces (Perez et al., 1990). Consequently, many more complex models have been developed to more accurately predict the diffuse irradiance of tilted surfaces from the solar data contained within weather files. Many of these separately predict the isotropic, circumsolar, and horizon components illustrated in Figure 9.4 and sum them to determine the total diffuse irradiance to the surface. One of these anisotropic models that has been implemented into many BPS tools is by Perez et al. (1990):

$$G_{solar\_diff \to e} = G_{solar\_diff \to \_}$$

$$\cdot \Big[(1 - F_1) \cdot f_{e \to sky\_dome} \qquad (9.13)$$

$$+ F_1 \cdot \frac{a}{b} + F_2 \cdot f_{e \to horizon} \Big]$$

The first term in the square brackets represents the isotropic component from the sky dome, while the second and third terms represent circumsolar and horizon brightening, respectively.

The second term of Equation 9.13 represents the circumsolar component, which is considered to have the same angle of incidence as the beam irradiance. The ratio $\frac{a}{b}$ is used to relate the irradiance to the tilted surface to that on the horizontal. a is a function of the angle of incidence with the surface under consideration, while b depends upon the solar altitude. For most sun positions this ratio assumes the value $\frac{\cos\theta}{\sin\eta_{alt}}$, the same factor as was seen in Equation 9.9.

Equation 9.13 contains two empirical factors. F_1 is the circumsolar brightening coefficient and F_2 is the horizon brightening coefficient. These are given as empirical functions of three parameters: the sun altitude, a sky *clearness* index, and a sky *brightness* index. The clearness index can be calculated from solar irradiance data contained within the weather file and indicates whether the sky is completely overcast, clear with low turbidity, or somewhere in between these extremes. The sky brightness index can also be calculated from weather file data and represents the opacity and thickness of the clouds. Because these indices span the full range of sky conditions, the Perez model is known as an *all-weather* model.

Parametric equations relate F_1 and F_2 to sun altitude, sky clearness, and sky brightness. Perez *et al.* regressed these equations using detailed measurements taken with pyrheliometers and pyranometers at six sites across the USA and at three sites in Europe to establish their empirical coefficients. Additional measurements from these sites and five others were then used to validate the model.

Many other anisotropic models exist. Some are cast in a form similar to Equation 9.13 to predict the isotropic, circumsolar, and horizon components, while others have different functional forms. Some are analytical in nature, but most have been regressed using measured data sets. You will gain a better understanding of the available models and their accuracies through one of this chapter's required readings.

It is important to realize that all BPS tools have a default model—whether that be isotropic or anisotropic—and some tools offer the user alternatives. You will explore which approaches are available in your chosen BPS tool during one of this chapter's simulation exercises.

9.5 GROUND-REFLECTED IRRADIANCE TO A BUILDING SURFACE

The ground will reflect a portion of the incoming solar irradiance, and some of this will be directed towards the surface under consideration. Although some types of ground surfaces reflect specularly, and others have directionally dependent properties, it is common practice for BPS tools to treat the ground as a diffuse reflector. With this approach, all beam and diffuse irradiance incident on the horizontal is assumed to be reflected equally in all directions (Figure 9.5):

$$G_{solar\_grd \to sky\_dome} = \rho_{grd} \cdot \left[G_{solar\_beam \to \_} + G_{solar\_diff \to \_} \right] \quad (9.14)$$

where ρ_{grd} is the total hemispherical reflectivity of the ground.

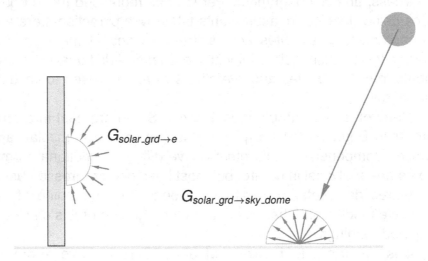

Figure 9.5: Ground-reflected irradiance

The ground-reflected irradiance to the building surface that is required by Equation 9.3 can be determined from the above quantity by making use of view factors and reciprocity, as was done for diffuse irradiance in Section 9.4:

$$G_{solar\_grd \to e} = f_{e \to grd} \cdot G_{solar\_grd \to sky\_dome} \quad (9.15)$$

The view factor from the surface to the ground can be determined by making use of Equation 9.12:

$$f_{e \to grd} = 1 - f_{e \to sky\_dome}$$

$$= \frac{1 - \cos \beta}{2} \tag{9.16}$$

Since $G_{solar\_beam \to \_}$ and $G_{solar\_diff \to \_}$ can be determined from the weather file data, once ρ_{grd} is established $G_{solar\_grd \to e}$ can be calculated with Equations 9.14 to 9.16.

However, establishing accurate ρ_{grd} values is a non-trivial task, as the reflectivity is strongly dependent upon the ground cover. For example, asphalt may have a ground reflectivity in the order of 0.05 to 0.15, whereas ρ_{grd} for a grass covered field could be in the range of 0.15 to 0.25, depending on the age, type, height, and moisture content of the grass. Freshly fallen snow may have a reflectivity as high as 0.95, but values of 0.5 or lower can occur for aged snow that has melted and refrozen. Cloud cover and the incidence angle of the solar irradiance are also determinants, so ρ_{grd} not only depends upon vegetation and seasonal effects, but can also vary throughout the day.

Notwithstanding this complexity, it is common practice for BPS tools to treat ρ_{grd} as time-invariant. Most tools use a default value of 0.2, but this can usually be overridden by the user, in some cases with monthly values. Some tools have optional methods to augment ρ_{grd} if the weather file indicates the presence of snow on the ground. However, most weather files have sparsely populated and sometimes inaccurate snow data, so such approaches can lead to unintended consequences. Although the commonly used EPW weather file format (see Section 8.5) includes a field for ρ_{grd}, this information is usually not included as this quantity is rarely measured.

Through one of this chapter's required readings you will learn about some models that have been developed to predict ρ_{grd} from other data contained in weather files. You will also explore the facilities available in your BPS tools for treating ground-reflected irradiance during one of this chapter's simulation exercises.

9.6   SHADING

Surrounding objects such as trees and other buildings can cast shadows which reduce the beam, diffuse, and ground-reflected solar irradiance to the surface. This is also the case for many architectural features, such as window fins and overhangs.

In BPS it is common practice to represent the shading effect of these objects by introducing *shading factors* to Equation 9.3:

$$q_{solar \to e} = \alpha_{solar,e} \cdot A_e \cdot \Big(\mathcal{F}_{beam} \cdot G_{solar\_beam \to e}$$

$$+ \mathcal{F}_{diff} \cdot G_{solar\_diff \to e} + \mathcal{F}_{grd} \cdot G_{solar\_grd \to e} \Big)$$

$$(9.17)$$

where $\mathcal{F}_{beam}$, $\mathcal{F}_{diff}$, and $\mathcal{F}_{grd}$ are the shading factors for beam, diffuse, and ground-reflected solar irradiance. The magnitude of these factors will vary in time and will depend upon the geometrical relationship between the shading objects and the surface under consideration, as well as upon sun position and sky conditions.

Most BPS tools possess models to calculate $\mathcal{F}_{beam}$. Some tools employ trigonometric relations, but these are limited to simple cases. Most BPS tools use a so-called *polygon clipping* method. With these, the shading object and the building surface under consideration undergo geometric transformations whereby they are projected onto a coplanar surface that is orthogonal to the sun beam. The degree of overlap between these projected surfaces determines the magnitude of $\mathcal{F}_{beam}$, with a value of 0 indicating complete shading. Various methods are used to determine the degree of overlap, some of which discretize the projected surfaces.

Some of the polygon clipping methods work well for simple configurations, but can produce significant prediction errors in more complex cases (de Almeida Rocha *et al.*, 2017). Pixel counting algorithms based on computer graphics rendering have been developed to address some of these shortcomings (Jones *et al.*, 2012), although few BPS tools support these capabilities.

Some BPS tools base the calculation of $G_{solar\_diff \to e}$ around the isotropic sky assumption, while others treat the sky as anisotropic and calculate the shading of the isotropic, circumsolar, and horizon components. It is common for many BPS tools to ignore the shading of ground-reflected irradiance, implicitly assuming $\mathcal{F}_{grd}$ to be 1.

The magnitude of the shading factors will vary in time due to the sun's apparent motion. To reduce the computational burden, many BPS tools do not calculate shading factors at each timestep of the simulation. In some cases they will be calculated only once per month by default, but often the processing frequency can be overridden by the user.

Some tools automatically consider self-shading effects wherein one element of the building shades others. However, this is not universally the case. With many BPS tools, shading is ignored by default and the user must take action to invoke optional capabilities and provide additional data.

You will explore the default approaches and optional methods provided by your BPS tool during one of this chapter's simulation exercises.

9.7 REQUIRED READING

Reading 9–A

Yang (2016) describes 26 models for predicting diffuse irradiance to tilted surfaces based upon horizontal measurements. They also evaluate the accuracy of these models through a comparison with measured data taken at four locations.

Read this article in its entirely and find answers to the following questions:

1. What is a transposition factor?
2. How many *families* of models are described in this article?
3. Which was the first model to consider circumsolar brightening?
4. Identify some of the models that separately predict the isotropic, circumsolar, and horizon components illustrated in Figure 9.4 to determine the total diffuse irradiance to the tilted surface.
5. What are the four sites of the measured data used in the val-

idation? What combinations of surface tilt and azimuth angles were examined?

6. Did they find that one model outperformed all others in every combination of site and surface orientation?

7. Which models were found to provide the best overall agreement with the measured data?

Reading 9–B

Dumitrascu and Beausoleil-Morrison (2020) describe a model for calculating ρ_{grd} that is based upon empirical measurements.

Read this article in its entirely and find answers to the following questions:

1. What is an albedometer?

2. This article includes a plot of the daily averaged ρ_{grd} that was measured at one site over the course of a full year. What was the range of daily averaged ρ_{grd} they observed?

3. At what times of the year did the lowest and highest values occur? Which month had the greatest variability in daily averaged ρ_{grd}? Why?

4. What are the five factors that were found to significantly influence ρ_{grd}?

5. Figures 7 and 8 of this article plot the global irradiance to a south-facing vertical surface on two winter days. What was the magnitude of the underprediction of this quantity when calculations were performed with $\rho_{grd} = 0.2$ (a common default value in BPS tools)?

9.8 SOURCES FOR FURTHER LEARNING

- Maestre *et al.* (2013) discuss the importance of accurately calculating shading. They review the main classifications of shading algorithms and explain why most BPS tools employ polygon clipping methods.

- The introduction of Cascone *et al.* (2011) provides a survey of methods that have been developed to calculate the shape of shadows cast on windows.

- de Almeida Rocha *et al.* (2017) present an experimental validation of polygon-clipping and pixel-counting shading models.

9.9 SIMULATION EXERCISES

Revert your BPS tool inputs to represent once again the *Base Case*
described in Section 1.9, including any refinements or corrections
you made following the previous exercises. Perform an annual simulation to produce a fresh set of *Base Case* results for use in the
following exercises.

Exercise 9–A

Solar absorptivity values were specified for the materials forming the external surfaces of all the opaque constructions in the
Base Case. Now reduce α_{solar} for the roof's asphalt shingles from
0.85 to 0.6 and perform another annual simulation using the same
timestep.

What impact does this change have upon the annual space
heating load? And upon the annual space cooling load? Are these
results in line with your expectations?

Contrast this to your results from Chapter 3 when you altered
the solar aborptivity of the floor's internal surface. Which change
had the greatest impact upon simulation predictions? Provide an
explanation.

Exercise 9–B

Revert to the *Base Case*.

What methods does your BPS tool provide for predicting the
distribution of diffuse solar irradiance? What is your tool's default
method? Which method did you use for simulating the *Base Case*?

Perform one or more additional simulations using alternate
methods offered by your BPS tool (e.g. isotropic sky model). What
impact does this have upon the annual space heating load? And
upon the annual space cooling load?

Are these results in line with your expectations based upon the material presented in Section 9.4 and in Reading 9–A? Which modelling option do you think is more appropriate?

Exercise 9–C

Revert to the *Base Case*.

The *Base Case* specifications provided little information on the ground surfaces surrounding the building. Did you assume a constant value for ρ_{grd}? If so, what value did you assume? If you used your BPS tool's default treatment, then describe what this treatment is.

Perform one or more simulations using alternate methods offered by your BPS tool for determining ρ_{grd}. Base your choices on what you learned from Reading 9–B.

Examine the predictions of solar irradiance incident upon the exterior surface of the window as a function of time on March 6 and contrast these predictions to those from the *Base Case*. What impact does this have upon the annual space heating load? And upon the annual space cooling load?

Think of some situations where it would be inappropriate to rely on your BPS tool's default approach.

Exercise 9–D

Revert to the *Base Case*.

There were no external objects that caused shading of the *Base Case*. Now perform another simulation which considers shading caused by a tree located south of the building. The tree has a diameter of 2 m and a height of 10 m. It is aligned with the window and is located 10 m south of the building.

Examine the solar irradiance incident upon the exterior surface of the window as a function of time on March 6 and contrast these predictions to those of the *Base Case*. What impact does this change have upon the annual space heating load? And upon the annual space cooling load? Are these results in line with your expectations?

Based upon these results, comment on the importance of considering shading effects by surrounding buildings, trees, and other objects. What are some situations in which the complexities of shading objects (e.g. size and types of trees, changing leaf cover) could be a significant factor?

9.10 CLOSING REMARKS

This chapter explained that it is common practice to calculate the global solar irradiance to external surfaces as the sum of three components: beam, diffuse, and ground-reflected. It showed how the beam component can be calculated from weather file data knowing the building's orientation and solar radiation geometry.

Through one of the required readings and the theory presented in this chapter, you learnt about the range of possible models that are used to calculate the diffuse component. Unlike the deterministic methods used to calculate beam irradiance, all these diffuse models make assumptions about the directional distribution of diffuse irradiance across the sky dome, some accounting for circumsolar and horizon brightening, while others not. You became aware of which diffuse models are available in your chosen BPS tool through the simulation exercises, and realized the actions required to invoke them.

This chapter and one of its required readings also explained that user choices on the treatment of ground reflectivity can significantly impact the calculation of the ground-reflected irradiance component. You saw first hand this impact through one of the simulation exercises.

An overview of the methods used to account for the shading of external surfaces by other building elements or surrounding objects was also provided, and you learnt how to invoke these facilities in your chosen BPS tool through one of the simulation exercises.

CHAPTER 10

Convection heat transfer at external surfaces

THIS chapter describes methods commonly used for calculating the convection heat transfer at external building surfaces.

Chapter learning objectives

1. Understand how convection heat transfer is represented in external surface energy balances.
2. Recognize BPS tool default and optional methods for determining convection coefficients.
3. Appreciate the complexities associated with establishing local wind speeds required by convection models.
4. Realize the sensitivity of simulation predictions to tool defaults and user choices.

10.1   MODELLING APPROACH

The $q_{conv,e \to oa}$ term in the external surface energy balance expressed by Equation 8.1 represents the convection heat transfer from surface e to the outdoor air. As with convection at internal surfaces (see Chapter 4) this term is commonly modelled using Newton's law of cooling:

$$q_{conv,e \to oa} = A_e \cdot h_{conv,e} \cdot (T_e - T_{oa}) \tag{10.1}$$

where A_e is the surface area (m^2), T_e the surface temperature (°C), and T_{oa} the outdoor air temperature (°C).

As we saw when we studied convection at internal surfaces, the coefficient $h_{conv,e}$ (W/m^2 K) characterizes the convection regime. Establishing appropriate values for $h_{conv,e}$ is key to accurately predicting $q_{conv,e \to oa}$.

The convection at the external surface may be driven by buoyancy or be caused by forced effects due to wind. In many cases both effects will be present. Consequently, the building's geometry, local wind speed and direction, and the surface and outdoor air temperatures can all influence $h_{conv,e}$. The geometry and proximity of surrounding buildings and topographical features can also impact local airflow patterns, and thus determine the local wind speed and direction over the surface under consideration.

Given this complexity, it is not surprising that many models have been developed to establish $h_{conv,e}$ coefficients.

10.2   CONVECTION CORRELATIONS

Most of the $h_{conv,e}$ models used by BPS tools have been derived from experimental data. For example, Walton (1983) developed the following relation for vertical surfaces:

$$h_{conv,e} = h_{natural} + h_{forced}$$

$$= 1.31 \cdot |T_e - T_{oa}|^{1/3} + 2.537 \cdot \mathcal{W}_f \cdot \mathcal{R}_f \cdot \left(\frac{P_L}{A}\right)^{1/2} \cdot V_e^{1/2} \tag{10.2}$$

The first term in this equation represents the contribution of natural convection effects. This was derived by Walton from the correlations of Churchill and Chu (1975), which are based on experiments conducted with small, isolated flat plates.

The second term of Equation 10.2 represents the contributions of forced convection effects caused by wind. It is based on data gathered by Sparrow *et al.* (1979) from experiments conducted with small, isolated flat plates in a wind tunnel. P_L represents the length of the wall's perimeter (m) while A is its area (m²). Walton included the $\mathcal{R}_f$ parameter (−) to augment (by up to two times) the convection coefficient for rough wall surfaces.

Walton's model includes two terms to account for the impact of the local wind speed. V_e in Equation 10.2 is the local wind speed at the height of surface e (m/s). The wind speed data contained in most weather files has been measured at a height of 10 m above the ground surface. An empirical relationship representing the shape of the wind's boundary layer over the ground surface is used to determine the local wind speed from the data contained in the weather file:

$$V_e = V_{wind} \cdot b_{wind} \cdot \left(\frac{H_e}{H_{wind}} \right)^{a_{wind}} \tag{10.3}$$

V_{wind} is the wind speed from the weather file (m/s), H_e is the height of surface e (m), and H_{wind} is the height at which the weather file's wind speed has been measured (m). a_{wind} and b_{wind} are unitless empirical constants that describe the shape of the atmospheric boundary layer. These depend upon the local terrain. The values of these coefficients are such that Equation 10.3 determines lower values of V_e in urban areas and city centres than in open and rural areas.

The unitless parameter W_f is the second term in Equation 10.2 that accounts for the local wind speed. When the wind direction in the weather file is within 100° of the surface normal, the surface is considered to be *windward* and W_f takes a value of 1. Otherwise the surface is considered to be *leeward* and W_f takes a value of 0.5. This approach assumes that the wind direction at the building site is the same as that recorded in the weather file, which essentially ignores the potential for the local terrain and surrounding buildings to alter wind directions.

In contrast to models such as Walton's that are based on

reduced-scale experiments, other approaches have been developed from full-scale experiments conducted on test buildings and occupied buildings. For example, the model of Yazdanian and Klems (1994) was developed from measurements taken at a low-rise test facility:

$$h_{conv,e} = \left\{ \left[\mathcal{C} \cdot \left| T_e - T_{oa} \right|^{1/3} \right]^2 + \left[\mathcal{A} \cdot V_{wind}^{\mathcal{B}} \right]^2 \right\}^{1/2} \qquad (10.4)$$

where $\mathcal{A}$, $\mathcal{B}$, and $\mathcal{C}$ are empirical constants that were regressed from the measured data. Separate values are given for windward and leeward surfaces.

Unlike Equation 10.2, Equation 10.4 uses the wind speed from the weather file, and as such the impact of boundary layer shape has already been accounted for in the empirical constants.

Many more empirical models like those of Walton and Yazdanian and Klems are available, as you will discover through one of this chapter's required readings.

Due to the impact of local airflow patterns on convection heat transfer, models based on full-scale measurements may only be strictly valid for buildings of similar geometry located in similar terrain. To address this limitation, numerous models for calculating $h_{conv,e}$ coefficients have recently been developed from detailed computational fluid dynamics simulations (e.g. Montazeri and Blocken, 2017). These models may be more broadly applicable as some include parameters related to building geometry, and as such are not limited to a single building shape.

10.3   LOCAL WIND SPEED AND USER OPTIONS

Section 4.5 discussed the approaches used for establishing convection coefficients at internal surfaces. Much the same options exist for external surfaces.

With the simplest option—the only one supported by some BPS tools—time-invariant $h_{conv,e}$ values are used. These may vary from surface to surface, and may be prescribed by the user or be defaulted by the BPS tool.

In some cases the user may opt to apply the BPS tool's default correlation to recalculate $h_{conv,e}$ values each timestep of the simulation. The tool may assign a default correlation based upon the surface's orientation (e.g. vertical).

Some BPS tools are limited to a single model to calculate $h_{conv,e}$ coefficients, while others may offer the user choices. In some cases the user can specify convection models surface-by-surface, and these are used to recalculate $h_{conv,e}$ values for each surface at each timestep of the simulation.

Chapter 8 pointed out that there can be considerable uncertainty in wind speed and direction data due to the hourly resolution of most weather files. Wind gusts and sudden changes in wind direction may be ignored (see Section 8.3). The user should be aware that this can substantially impact the $h_{conv,e}$ values predicted by Equations 10.2 and 10.4 or other such models.

Some BPS tools apply relations like Equation 10.3 to account for the local terrain while other tools ignore this effect. Likewise, some tools account for surface elevation when calculating the local wind speed to use in models, while others do not. Some BPS tools may account for surface roughness effects (and require additional user inputs) while others do not.

You will learn more about which models are supported by your chosen BPS tool and how it treats effects such as local terrain and surface elevation during this chapter's simulation exercises.

10.4 REQUIRED READING

Reading 10–A

Mirsadeghi *et al.* (2013) describe the empirical correlations that have been developed for calculating $h_{conv,e}$ coefficients for buildings. They also discuss how these correlations have been implemented into some common BPS tools.

Read the first two sections of this article and answer the following:

1. What type of *reduced-scale* experiments were used to develop the first group of correlations listed in the article's Table 12?

2. As the article explains, the McAdams model has been imple-

mented into a number of BPS tools. What is the source of the experimental data that were used to develop this model?

3. Describe some of the full-scale experiments that have been used to derive convection models.

4. What is the definition of *free stream wind speed*? And of *local wind speed*? Some of the models described in the article require knowledge of these quantities. How are they determined from the wind speed data contained in the weather file?

5. What does surface *roughness* mean? What impact does roughness tend to have on $h_{conv,e}$ coefficients?

6. Which of the BPS tools considered by this article allow the user to choose between numerous empirical models for calculating $h_{conv,e}$ coefficients?

Reading 10–B

Iousef *et al.* (2019) investigate the impact of the choice of external convection model upon simulation predictions. They conducted simulations using several convection models and examined buildings with various geometries and insulation levels.

Find answers to the following based upon their results:

1. Are simulation predictions of well-insulated buildings or poorly insulated buildings more sensitive to the choice of convection model? Why?

2. Which simulation predictions are more sensitive to the choice of convection model: annual space heating load or annual cooling load? Why?

3. Are simulation predictions of low-rise or high-rise buildings more sensitive to the choice of convection model? Why?

10.5   SOURCES FOR FURTHER LEARNING

- Montazeri and Blocken (2017) and Montazeri and Blocken (2018) present correlations for calculating $h_{conv,e}$ that consider wind speed and direction, as well as the building's width and height. These correlations are derived from detailed computational fluid dynamics simulations that have been validated using wind tunnel measurements.

10.6 SIMULATION EXERCISES

Revert your BPS tool inputs to represent once again the *Base Case* described in Section 1.9, including any refinements or corrections you made following the previous exercises. Perform an annual simulation to produce a fresh set of *Base Case* results for use in the following exercises.

Exercise 10–A

How did you treat external surface convection in your *Base Case*? Did you use your BPS tool's default approach? This is likely the case if you did not take action to override your tool's default method. Consult your BPS tool's help file or technical documentation to determine its default approach for establishing $h_{conv,e}$ coefficients.

If you used your BPS tool's default method, then configure your tool to now impose time-invariant $h_{conv,e}$ coefficients of $15\,W/m^2\,K$ at all external surfaces and perform another annual simulation using the same timestep.

If you did not use the default method provided by your BPS tool for the *Base Case*, then configure your tool to now employ its default method and perform another annual simulation using the same timestep.

Compare the results of the two simulations. What impact does this change have upon the annual space heating load? And upon the annual space cooling load? Are these results consistent with the observations presented in Reading 10–B?

Exercise 10–B

Does your BPS tool provide optional models for calculating $h_{conv,e}$ coefficients? If so, which of the correlations listed in Reading 10–A does your tool support? Does it support additional correlations as well?

Configure your BPS tool to apply one or more of its optional correlations and perform some additional simulations. What impact do these changes have upon the annual space heating load? And upon the annual space cooling load?

Exercise 10–C

Does your BPS tool support the Walton (1983) model described in Section 10.2, or another model that includes the roughness parameter $\mathcal{R}_f$? If so, configure your tool to apply this model and perform two simulations: one with all the exterior walls set to *smooth* and the other with them set to *very rough*.

What impact does the $\mathcal{R}_f$ parameter have on the annual space heating load? And upon the annual space cooling load?

What are possible sources of information that can be used to determine $\mathcal{R}_f$ values for materials?

Exercise 10–D

Revert to your *Base Case* and invoke your tool's default method for external surface convection.

Section 10.2 explained the uncertainty in establishing local wind speeds used to calculate $h_{conv,e}$ coefficients. To assess the sensitivity to this, reduce the weather file's wind speed by 50 % for each hour of the year. This can be accomplished by manipulating the CWEC-2016 weather file using the Elements software mentioned in Exercise 8–C.

What impact does this change have upon the annual space heating load? And upon the annual space cooling load?

10.7 CLOSING REMARKS

This chapter explained that it is common practice for BPS tools to represent the convection heat transfer between external surfaces and the outdoor air using Newton's law of cooling. This approach requires the determination of convection coefficients for each external surface, and these can vary with each timestep of the simulation.

Through the theory presented in this chapter and one of the required readings, you became aware of the numerous models used in the BPS field for calculating the required convection coefficients. Most of these consider forced convection effects caused by wind, and you became aware of the complexity of estimating local wind speeds from the data contained within weather files.

You discovered your BPS tool's default and optional methods for treating convection at external surfaces by conducting the simulation exercises. These also helped you appreciate the impact that user choices (including relying on default methods) can have upon simulation predictions.

CHAPTER 11

Longwave radiation exchange at external surfaces

T HIS chapter describes methods for treating longwave radiation between external building surfaces and the exterior environment, including the sky.

Chapter learning objectives

1. Understand the methods used to calculate longwave radiation exchange between external surfaces and the exterior environment, and their inherent approximations and simplifications.

2. Understand the methods used to establish radiation view factors between external surfaces and environment, and how these can be influenced by user inputs.

3. Realize the uncertainties associated with models that are used to estimate the sky temperature.

4. Realize the sensitivity of simulation predictions to user choices and inputs.

11.1 MODELLING APPROACH

The net exchange of longwave radiation from surface e to the exterior environment is represented in the external surface energy balance of Equation 8.1 by the term $\sum_{env} q_{lw,e \to env}$.

A summation is used because the external surface exchanges longwave radiation with all entities that radiate in the longwave spectrum and which can be viewed by the surface. This includes the surface of the surrounding ground, neighbouring buildings, nearby structures, and objects such as trees. Atmospheric gases such as water vapour, carbon dioxide, and ozone can also exchange longwave radiation with external building surfaces, as can atmospheric aerosols and clouds. And when the sky is not obscured by clouds the surface can exchange longwave radiation with deep space.

In reality the $\sum_{env} q_{lw,e \to env}$ summation should include many terms because each of these entities participating in the longwave radiation exchange is at a different temperature. For example, the temperature of deep space is $\sim 3\,\text{K}$, whereas water vapour molecules near the ground surface will be close to the ambient air temperature, while those at higher elevations in the atmosphere will be much colder. However, to render the problem manageable it is common practice for BPS tools to represent the longwave radiation exchange from external surfaces as the summation of the three components illustrated in Figure 11.1:

$$\sum_{env} q_{lw,e \to env} = q_{lw,e \to sky} + q_{lw,e \to grd} + q_{lw,e \to obj} \qquad (11.1)$$

The *sky* term represents longwave radiation exchange with all the participating entities in the earth's atmosphere and beyond: atmospheric gases, aerosols, clouds, and deep space. An effective sky temperature, T_{sky}, is used to represent a uniform sky that considers the combined effect of the radiation exchange with each participating entity. T_{sky} can be as warm as $15\,°\text{C}$ under warm and cloudy conditions, and colder than $-40\,°\text{C}$ when it is cold and the sky clear.

Likewise, the surface of the surrounding ground is treated as being at a uniform temperature of T_{grd}, and all surrounding objects

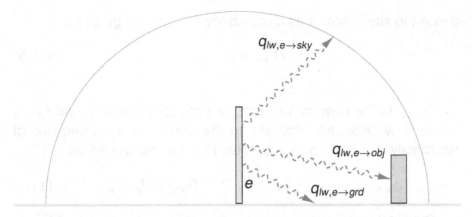

Figure 11.1: Longwave radiation exchange between an external surface and the exterior environment

such as buildings and trees are treated as being at a common temperature of T_{obj}. These tend to be warmer than T_{sky}, and as such, the $q_{lw,e\rightarrow sky}$ term usually dominates in Equation 11.1.

11.2 CALCULATING LONGWAVE EXCHANGE

Chapter 5 explained the methods used to calculate longwave radiation exchange between internal building surfaces. These principles apply equally here, although in this case the external surface is exchanging longwave radiation with an environment that can be considered as an infinitely large enclosure.

Consider the first term of Equation 11.1. Due to the infinite enclosure assumption, the sky can be treated as a blackbody ($\epsilon_{sky} = 1$) since any emission from surface e directed towards the sky will eventually be absorbed rather than reflected back to the surface (refer to Section 5.3). With this, the net longwave radiation exchange from the surface to the sky can be written as:

$$q_{lw,e\rightarrow sky} = \left(\epsilon_{lw,e} \cdot \sigma \cdot A_e \cdot T_e^4\right) \cdot f_{e\rightarrow sky}$$
$$- \epsilon_{lw,e} \cdot \left(A_{sky} \cdot G_{sky\downarrow}\right) \cdot f_{sky\rightarrow e} \tag{11.2}$$

$G_{sky\downarrow}$ is the longwave irradiance emitted by the sky (W/m^2) and is commonly referred to as the *downwelling* longwave irradiance.

Since the sky behaves as a blackbody this can be given as:

$$G_{sky\downarrow} = \sigma \cdot T_{sky}^4 \qquad (11.3)$$

$f_{e\rightarrow sky}$ is the view factor from the surface to the sky, and $f_{sky\rightarrow e}$ is the view factor from the sky to the surface. By making use of reciprocity (Equation 5.8), Equation 11.2 can be written as:

$$q_{lw,e\rightarrow sky} = \epsilon_{lw,e} \cdot A_e \cdot \left(\sigma \cdot T_e^4 - G_{sky\downarrow}\right) \cdot f_{e\rightarrow sky} \qquad (11.4)$$

Substituting the expression for downwelling irradiance of Equation 11.3 into the above leads to:

$$q_{lw,e\rightarrow sky} = \epsilon_{lw,e} \cdot \sigma \cdot A_e \cdot \left(T_e^4 - T_{sky}^4\right) \cdot f_{e\rightarrow sky} \qquad (11.5)$$

The same blackbody approximation is usually extended to the other two terms of Equation 11.1, even though it is strictly less true in the case of surrounding objects. With this, Equation 11.1 can be expressed as:

$$\sum_{env} q_{lw,e\rightarrow env} = \epsilon_{lw,e} \cdot \sigma \cdot A_e \cdot \Bigg[\left(T_e^4 - T_{sky}^4\right) \cdot f_{e\rightarrow sky}$$
$$+ \left(T_e^4 - T_{grd}^4\right) \cdot f_{e\rightarrow grd} + \left(T_e^4 - T_{obj}^4\right) \cdot f_{e\rightarrow obj} \Bigg] \qquad (11.6)$$

where the three view factors sum to unity.

11.3   EFFECTIVE TEMPERATURES AND VIEW FACTORS

The effective temperatures T_{sky}, T_{grd}, and T_{obj} and the view factors between the external surface and these entities must be established in order to solve Equation 11.6.

Some BPS tools attempt to calculate T_{grd} by forming an energy balance at the ground surface that considers solar absorption, conduction through the soil, convection to the ambient air, and infrared radiation to the sky. In this way T_{grd} can evolve during the simulation. However, such rigour is rare, and important effects such as snow cover and snow melting are usually neglected. A common simplification is to assume that the ground surface is at the ambient air temperature: $T_{grd} = T_{oa}$.

Some BPS tools can explicitly calculate the temperature of the external surfaces of surrounding buildings, if the user has described these in sufficient detail, but this practice is rare. Some other tools assume that surrounding buildings are at a similar temperature, perhaps by averaging the temperatures of the exterior surfaces of the building under consideration. A more common approach is to assume that surrounding buildings and objects are at the ambient air temperature: $T_{obj} = T_{oa}$.

During this chapter's simulation exercises you will explore how your chosen BPS tool determines T_{grd} and T_{obj}.

Many BPS tools use the geometry data provided by the user to establish the view factors using the methods already described in Chapter 9 (refer to Equations 9.12 and 9.16). With this approach a horizontal flat roof would have a greater value of $f_{e \to sky}$ than a vertical wall, for example.

Some BPS tools apply user-supplied contextual information— such as whether the building is located in a dense urban area or at a rural site—to help establish appropriate view factors. For example, the vertical surfaces of a building located in a densely populated urban area would have a greater $f_{e \to obj}$ and a lower $f_{e \to sky}$ than the vertical surfaces of a building located on a rural site. Some tools allow the user to specify their own view factors.

Establishing appropriate view factors can have a significant impact upon Equation 11.6 because, in general, T_{sky} can be significantly colder than T_{grd} and T_{obj}. All BPS tools employ some default method to determine the view factors and the user should be aware of these assumptions and the optional methods available. You will explore which methods are available in your chosen BPS tool during this chapter's simulation exercises, and you will examine the impact view factors can have upon simulation predictions.

The final term that needs to be established to evaluate Equation 11.6 is T_{sky}.

11.4 SKY TEMPERATURE

It is possible to measure the downwelling longwave irradiance using a pyrgeometer, a device resembling a pyranometer that is used to measure solar irradiance. If $G_{sky\downarrow}$ were directly measured in this way and made available in the weather file, then it could be directly used in Equation 11.4 to calculate $q_{lw,e\rightarrow sky}$.

However, very few weather stations possess pyrgeometers, so this measurement is not usually available. If $G_{sky\downarrow}$ is included in the weather file (the EPW format includes a record for it) then some BPS tools will use these data to solve Equation 11.4, but most tools will ignore it in favour of the models presented in this section. Also, when $G_{sky\downarrow}$ has been included in the weather file it has usually been approximated by the same type of models.

The common practice in the BPS field is to calculate $q_{lw,e\rightarrow sky}$ using Equation 11.5 and to employ a model to estimate T_{sky} from commonly available weather parameters. Most models relate T_{sky} to the ambient air temperature[i]. In general T_{sky} will be (significantly) colder than T_{oa}. Therefore, a fictitious quantity known as the *sky emissivity* is introduced to represent the blackbody emission of the sky by a grey surface emitting at the (higher) temperature of T_{oa}:

$$G_{sky\downarrow} = \epsilon_{sky} \cdot \sigma \cdot T_{oa}^4 \qquad (11.7)$$

where ϵ_{sky} is the fictitious sky emissivity, which has a value between 0 and 1.

The required value of T_{sky} can be determined by equating Equations 11.3 and 11.7:

$$T_{sky} = \epsilon_{sky}^{1/4} \cdot T_{oa} \qquad (11.8)$$

Many models have been developed to estimate ϵ_{sky}, most of which are based upon empirical data gathered in a single location, or from multiple sites within the same country. One such model that is used in a number of BPS tools is that of Martin and Berdahl (1984). It is based upon several years of pyrgeometer measurements that were taken at six locations across the southern USA.

---

[i]You will see reference to the *screening* temperature in the meteorology and climatology literature, where screen refers to a Stevenson screen. This is equivalent to the T_{oa} in weather files.

Martin and Berdahl noted that under clear-sky conditions ϵ_{sky} could be related to the humidity of the ground-level air (as measured by its dew point temperature). Additionally, they noted that ϵ_{sky} tended to be slightly higher at night. Based upon these observations they developed a functional form for a model and calibrated its constants using their measurements:

$$\epsilon_{sky} = 0.711 + 0.56 \cdot \left(\frac{T_{dp}}{100}\right) + 0.73 \cdot \left(\frac{T_{dp}}{100}\right)^2$$
$$+ 0.013 \cdot \cos\left(\frac{2\pi t}{24 \cdot 3600}\right) + 0.000\,12 \cdot \left(\frac{P_{atm}}{100\,000} - 100\,000\right)$$
(11.9)

T_{dp} is the ground-level dew point temperature (°C). It is either directly input from the weather file, or the BPS tool calculates it from other humidity parameters. P_{atm} is the atmospheric pressure (Pa) which is usually extracted from the weather file.

The impact of humidity can be seen by examining Equations 11.5, 11.8, and 11.9. A higher T_{dp} leads to a larger T_{sky} and therefore to a lower $q_{lw,e \to sky}$.

The fourth term in Equation 11.9 accounts for diurnal variations (t is the time since midnight in seconds) and leads to slightly higher ϵ_{sky} values at night. Martin and Berdahl included the last term in Equation 11.9 based upon the earlier work of Staley and Jurica (1972), which showed that ϵ_{sky} decreases with site elevation due to a decrease in optical depth of the atmosphere.

Many of the models used in BPS tools account for the influence of clouds since low-level opaque clouds increase $G_{sky\downarrow}$. One such model is that of Clark and Allen (1978), which is based upon measurements taken at a single location in the USA. Measurements were taken with a net radiometer that measured the difference between $G_{sky\downarrow}$ and upwelling longwave radiation from a pond of water. The upwelling radiation included both the emission from the water as well as the reflected $G_{sky\downarrow}$. By estimating ϵ_{lw} of the water they were able to calculate ϵ_{sky} from the radiometer measurements as well as measurements of the temperature of the ambient air and of the pond, although the measurement uncertainties associated with this approach would be significant.

Based upon this, they derived the following model:

$$\epsilon_{sky} = \left[0.787 + 0.764 \cdot \ln \left(\frac{T_{dp}}{273} \right) \right] \cdot C_a \qquad (11.10)$$

where T_{dp} is measured in K in this case.

The term enclosed in the square brackets represents ϵ_{sky} under clear sky conditions, and is thus comparable to the Martin and Berdahl model of Equation 11.9. C_a is a scalar multiplier that accounts for the impact of clouds and is given by:

$$C_a = 1 + 0.0224 \cdot N - 0.0035 \cdot N^2 + 0.00028 \cdot N^3 \qquad (11.11)$$

The parameter N in Equation 11.11 represents the degree of cloud cover. This ranges from 0 to 10 and represents the amount of the sky dome that is obscured by opaque clouds, where a value of 5 indicates that 50 % of the sky dome is obscured and a value of 10 indicates complete obscuration. This is a parameter that is commonly observed and available in weather files. Again, the parameters of this equation were calibrated from the measurements taken by Clark and Allen at a single location.

Many other ϵ_{sky} models exist with different functional forms, as you will discover through this chapter's required readings. All are empirical in nature, having been calibrated using data gathered at particular locations.

11.5 REQUIRED READING

Reading 11–A

Evangelisti *et al.* (2019) review a number of models that have been developed to estimate the effective sky temperature.

Read Section 3 of this article and find answers to the following questions:

1. How many of the clear-sky models have been calibrated with empirical data gathered in multiple countries?
2. This article differentiates between *direct* models and those based upon emissivity correlations. What is the difference between these two classes of model?

3. Most of the clear-sky models are functions of humidity to consider the important role that water vapour in the earth's atmosphere plays in longwave radiation exchange. What are the various parameters that are used to represent humidity in the models? Why do you think different models use different measures of humidity?

4. Which of the clear-sky models do not consider humidity? Why don't these consider humidity?

5. Why are there fewer cloudy-sky models than clear-sky models?

6. Most of the cloudy-sky models include a parameter to indicate the cloud extent. What are the various measures of cloudiness that are used?

Reading 11–B

Li et al. (2017) empirically validate a number of longwave sky radiation models. Specifically, they compare pyrgeometer measurements of downwelling longwave irradiance to model predictions. The measurements were taken at one-minute intervals at seven sites throughout the USA. They also recalibrate these models using these new measurements.

Read Section 4 of this article and find answers to the following:

1. Which of the models tested by Li et al. were described in Reading 11–A?

2. Which of the models in their original form were biased to underpredict the measured downwelling longwave irradiance? And which were biased to overpredict?

3. Which of the models in their original form had relative root mean square errors of 10 % or greater?

4. Why did the predictive accuracy of all the models improve after recalibration?

5. Even after recalibration, the Swinbank and Jackson-Ruchstuhl models still had significant root mean square errors. What do the functional forms of these two models have in common?

11.6   SOURCES FOR FURTHER LEARNING

- Čekon (2015) compares measured heat transfer rates to the predictions of several longwave radiation models under clear-sky conditions.
- Zhang *et al.* (2017) compare three T_{sky} models used by BPS tools to measured data and assess the impact of algorithmic differences on simulation predictions.
- Section 5 of Evangelisti *et al.* (2019) compares the sky emissivity and sky temperature predictions of various locations with climates varying from tropical, to mild, to snowy.
- Section 5 of Li *et al.* (2017) presents a new model for predicting the downwelling longwave irradiance from the sky under clear-sky and cloudy-sky conditions.

11.7   SIMULATION EXERCISES

Revert your BPS tool inputs to represent once again the *Base Case* described in Section 1.9, including any refinements or corrections you made following the previous exercises. Perform an annual simulation to produce a fresh set of *Base Case* results for use in the following exercises.

Exercise 11–A

Longwave emissivity values were specified for the materials forming the external surfaces of all the opaque envelope constructions in the *Base Case*. Now reduce ϵ_{lw} of the wall's cedar and of the roof's asphalt shingles from 0.9 to 0.6 and perform another annual simulation using the same timestep. If your BPS tool does not allow you to modify a surface's ϵ_{lw} then consult its technical documentation to determine why.

What impact does this change have upon the annual space heating load? And upon the annual space cooling load? Are these results in line with your expectations?

Contrast this to your results from Chapter 5 when you altered ϵ_{lw} of the internal surfaces. Which change had the greatest impact upon simulation predictions? Provide an explanation.

What are possible sources of information that can be used to determine ϵ_{lw} values for materials? Describe a situation in which accurately establishing a material's longwave emissivity may have an important impact upon simulation predictions.

Exercise 11–B

Undo the change made in Exercise 11–A.

Consult your BPS tool's help file or technical documentation to determine its default approach for establishing radiation view factors between external surfaces and the exterior environment. (Refer to Section 11.3.) How did you configure your Base Case to determine these view factors? If you did not use the default method provided by your BPS tool, then configure your tool to employ its default method and perform another annual simulation using the same timestep.

Now configure your tool to alter the radiation view factors for the external walls and the roof: increase $f_{e \rightarrow sky}$ and decrease the others. This scenario could represent, for example, a more rural setting for the building. Perform additional annual simulations using the same timestep for various scenarios.

What impact do these changes have upon the annual space heating load? And upon the annual space cooling load? Are these results in line with your expectations?

If your BPS tool does not allow you to modify these view factors then consult its technical documentation to determine why.

Exercise 11–C

Undo the change made in Exercise 11–B.

Consult your BPS tool's help file or technical documentation to determine its default approach for calculating T_{sky}. (Refer to Section 11.4.) How did you configure your Base Case to calculate T_{sky}? If you did not use the default method provided by your BPS tool, then configure your tool to now employ its default method and perform another annual simulation using the same timestep.

Does your BPS tool provide optional methods for calculating

T_{sky}? If so, which of the models described in Reading 11–A does your tool support? Does it support additional methods?

Configure your BPS tool to apply one or more of its optional T_{sky} models and perform some additional simulations. What impact do these changes have upon the annual space heating load? And upon the annual space cooling load?

If your tool does not provide optional T_{sky} models, then manually calculate T_{sky} for clear sky conditions using the two models expressed by Equations 11.9 and 11.10, making any necessary assumptions. Do this for a cool winter day when T_{oa} = −11 °C and T_{dp} = −20 °C. And then again for a milder day when T_{oa} = 10 °C and T_{dp} = 0 °C.

How do the T_{sky} predictions differ between these two models? What impact do you think this would have upon $q_{lw,e \rightarrow sky}$ and upon the external surface energy balance?

Exercise 11–D

Undo the change made in Exercise 11–C.

Consult your BPS tool's help file or technical documentation to determine its default approachs for calculating T_{grd} and T_{obj}. (Refer to Section 11.3.) How did you configure your *Base Case* to calculate T_{grd} and T_{obj}? If you did not use the default methods provided by your BPS tool, then configure your tool to now employ its default methods and perform another annual simulation using the same timestep.

Does your BPS tool provide optional methods for calculating T_{grd} and T_{obj}? Describe the default and optional methods.

Configure your BPS tool to apply one or more of its optional T_{grd} and T_{obj} models and perform some additional simulations. What impact do these changes have upon the annual space heating load? And upon the annual space cooling load?

11.8 CLOSING REMARKS

External building surfaces exchange longwave radiation with the surrounding ground, neighbouring buildings and structures, atmospheric gases and aerosols, clouds, and deep space. The media in the atmosphere and deep space dominate this radiation exchange because in general they are at substantially colder temperatures.

This chapter explained that it is common practice for BPS tools to represent the sky as uniform at an effective temperature to characterize the combined effect of the radiation exchange to the media in the atmosphere and deep space. Through the theory presented in this chapter and the required readings you became aware of the models that are used by BPS tools for estimating the effective sky temperature. Most of these operate with data commonly available in weather files, such as ground-level humidity and cloud cover. You became aware of the predictive differences between these models through one of the required readings.

Through the simulation exercises you discovered which methods are available in your chosen BPS tool for estimating the effective sky temperature and for establishing radiation view factors. And you realized the impact of optional methods and user choices upon simulation predictions.

CHAPTER 12

Heat transfer to the ground

T HIS chapter describes the wide-ranging approaches that are available within BPS tools for treating heat transfer at external surfaces in contact with the ground.

Chapter learning objectives

1. Understand the modelling options for resolving this important heat transfer path.
2. Realize the implications of simplified approaches.
3. Appreciate the uncertainty associated with prescribing soil thermophysical properties and boundary conditions.
4. Learn how to configure BPS tools to apply optional models.

12.1 MODELLING GROUND HEAT TRANSFER

The $q_{e \to grd}$ term of Equation 8.1 represents the rate of heat transfer from the exterior surface to the surrounding soil. This term is only relevant when the exterior surface under consideration is below-grade, such as with floor slabs and foundation walls.

This heat transfer is complex, being both transient and three-dimensional. In reality, heat conduction, convection, and radiation processes occur within the soil underneath and adjacent to the foundation, and these thermal processes are coupled to mass transfer processes that include changes of phase. Water will migrate through the soil following rain events and these migration patterns will be influenced by building drainage systems. The water may also freeze and thaw and these phase-change processes will be influenced by heat transfer from the building.

The heat and mass transfer processes occurring at the ground surface are also complex, and include factors such as infrared radiation to the sky, solar absorption, convection to the ambient air, water evaporation, and snow accumulation and melting.

Most models that consider the thermal coupling between the building and the ground neglect many of these effects for pragmatic reasons, and instead consider heat transfer through the ground as a conduction-only process where an effective conductivity approximately represents all of these processes. In such cases the effective conductivity should be dependent upon the soil moisture content, although this is rarely represented in models.

There are other challenges in modelling the coupling between the building and the ground. The thermal response of the ground is much slower than that of building elements, which can have a significant impact upon simulation run time. Large computational domains may be required if ground boundary conditions are placed sufficiently far from buildings such that they are not influenced by the building's presence. In such situations accuracy may have to be sacrificed for the sake of computational efficiency. Finally, weather files typically contain little or no information on ground temperatures.

There is a wide spectrum of methods used in the BPS field for

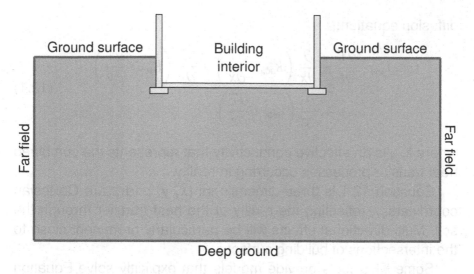

Figure 12.1: Calculation domain for detailed numerical ground models (soil indicated by ■)

treating heat transfer to the ground. Some BPS tools offer multiple detailed modelling options, while others lack models and instead rely upon the user to somehow configure the tool to approximate $q_{e \rightarrow grd}$. For this reason, the calculation of heat transfer to the ground remains one of the most significant sources of uncertainty in BPS.

12.2   DETAILED NUMERICAL METHODS

A building foundation and its surrounding soil are illustrated in Figure 12.1. This shows three external building surfaces (two walls and a floor) and the structural footing supporting them. The soil surrounding these surfaces is bounded by several planes: horizontal planes at the ground surface and at some depth below-grade, and vertical planes (two can be seen in the figure) at some distance from the foundation.

As explained above, it is common practice to represent heat transfer in the soil as a conduction-only process. Given this simplification, heat transfer in the soil can be described with the heat diffusion equation:

$$
(\rho c_P)_{soil} \cdot \frac{\partial T}{\partial t} = \frac{\partial}{\partial x}\left(k_{soil} \cdot \frac{\partial T}{\partial x}\right) + \frac{\partial}{\partial y}\left(k_{soil} \cdot \frac{\partial T}{\partial y}\right)
$$
$$
+ \frac{\partial}{\partial z}\left(k_{soil} \cdot \frac{\partial T}{\partial z}\right)
$$

(12.1)

where k_{soil} is an effective conductivity that represents the combined heat transfer processes occurring in reality.

Equation 12.1 is three-dimensional (x, y, and z are Cartesian coordinates), reflecting the reality of the heat transfer through the soil. Multi-directional effects will be particularly prominent close to the intersections of building surfaces.

Some BPS tools provide models that explicitly solve Equation 12.1 using numerical methods. Finite-difference methods are most commonly employed, although other approaches are possible. This involves subdividing the soil within the calculation domain into a number (perhaps hundreds or thousands) of volumes, and for each volume approximating the partial derivatives of Equation 12.1 with algebraic expressions. This produces a set of coupled algebraic equations (one for each volume) that can then be solved iteratively once closed with initial and boundary conditions. (You will learn about these numerical methods in Chapter 13.)

The user must provide a significant amount of information when such a model is invoked, such as soil properties (ρ, c_P, and k for Equation 12.1) and the number and dimensions of the volumes used to spatially discretize the soil domain. Conditions must also be prescribed at each of the boundaries illustrated in Figure 12.1.

In some cases the user may have to specify the depth of the *deep ground* boundary and the length of the *far field* boundaries from the building. Sometimes the deep ground boundary is treated as being isothermal (and therefore requiring a prescribed temperature from the user), and sometimes it is treated as adiabatic. Likewise, the user often has to choose the boundary condition treatment for the far field. Common options are adiabatic or a user-selected function prescribing undisturbed ground temperatures as a function of depth.

Some BPS tools will consider solar radiation absorption, convection, and longwave radiation heat transfer at the *ground surface* boundary using the methods described in Chapters 9, 10, and 11. The boundary conditions at the interface between the building surfaces and the soil can either be prescribed by imposing the T_e temperatures determined by the BPS tool's solver (refer to Section 2.8), or the numerical domain could be extended to represent the building constructions and then boundary conditions prescribed at the internal surfaces.

The discretized energy balances representing Equation 12.1 for each volume can then be solved subject to this set of boundary conditions. The resulting solution of the soil's temperature field can be used to determine $q_{e \rightarrow grd}$, or the calculated temperature of the soil adjacent to the building external surface can be imposed as a boundary condition on the model used to solve $q_{cond, m \rightarrow e}$ of Equation 8.1. (These models will be treated in Chapter 13.)

Because the large mass of soil contained within the computational domain can store a significant amount of energy, the treatment of initial conditions can have an important impact upon the computational time required to exercise such detailed numerical models. Some BPS tools include techniques for mitigating this (Kruis and Krarti, 2015). As the computational burden of explicitly representing heat transfer in three dimensions can also be excessive, methods have been developed to use two-dimensional calculations to estimate the heat transfer. You will learn about these through one of this chapter's required readings.

These detailed numerical methods—which are only available in a few BPS tools—offer the potential for the greatest accuracy (if used correctly) since they consider the storage of energy in the surrounding soil and details of the thermal interaction between the building and the soil. However, they demand considerable effort from the user and require significant computational resources.

You will discover whether your chosen BPS tool supports a detailed numerical method during one of this chapter's simulation exercises. If it does, you will learn the types of inputs that it requires and explore how to configure it.

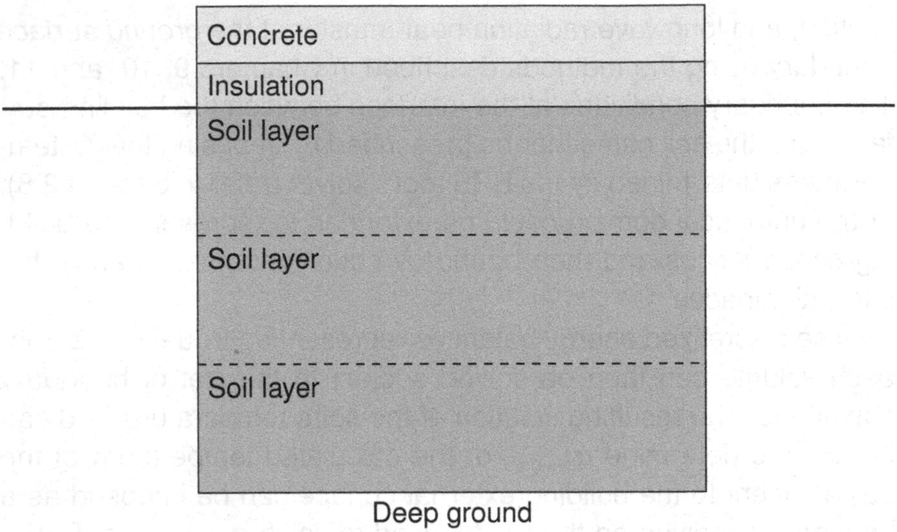

Figure 12.2: Adding soil layers to a below-grade envelope construction

12.3 ADDING SOIL TO BELOW-GRADE ENVELOPE CONSTRUCTIONS

This is another method for predicting $q_{e \to grd}$. Although less accurate than the detailed numerical methods described in the previous section, this technique can be applied in most BPS tools. With this, the user adds additional material layers to the construction to represent the surrounding soil, and the BPS tool's model for calculating transient heat transfer through opaque building envelope assemblies is employed.

Figure 12.2 illustrates a slab-on-grade construction similar to that of the *Base Case*. The floor's concrete and insulation layer are shown in the figure. Three layers of soil are added to the construction. The number and thickness of the layers is at the user's discretion.

The user must provide thermophysical properties (ρ, c_P, and k) to represent the soil. The user must also prescribe the boundary condition at the *deep ground* location shown in the figure. It is common to treat this boundary with a prescribed, constant temperature. Although there is no consensus on what this temperature should be, some tools suggest that the annually averaged ambient air temperature is a good approximation if the soil layers are made sufficiently thick (perhaps 5 m or more). However, as some BPS tools are unable to treat assemblies with this thickness or mass, the user is often faced with finding a compromise.

The BPS tool's model for calculating transient heat transfer through opaque envelope assemblies (the subject of Chapter 13) is then used to calculate the heat transfer through the combined assembly of the construction and soil. As such, $q_{e \rightarrow grd}$ is set to zero since the $q_{cond,m \rightarrow e}$ term of Equation 8.1 will represent the combined influence of the construction and the soil.

This method considers thermal storage of the soil surrounding the construction, although the spatial resolution of the solution is likely coarser than can be achieved with the detailed numerical methods described in the previous section. However, since heat transfer through the soil is treated as one-dimensional (most BPS tools treat heat transfer through opaque envelope assemblies one-dimensionally, as will soon be seen in Chapter 13) important multi-dimensional effects are ignored. The consequence of this is usually an underestimation of heat transfer from the external surface to the soil. Some tools use methods to augment the calculations to compensate for this.

You will exercise this modelling approach during one of this chapter's simulation exercises.

12.4 REGRESSION AND ANALYTICAL APPROACHES

A number of regression-based methods derived from detailed numerical simulations have been developed. Most of these are based upon the same governing equation and boundary condition treatments solved by the detailed numerical methods described in Section 12.2.

However, rather than solving the energy balances for the soil volumes within the BPS tool, these detailed calculations are performed externally to the BPS tool. Typically the authors of these methods have performed a large number (sometimes exceeding 100 000) numerical simulations to span a wide range of expected parameters, such as foundation geometry, thermophysical properties, and insulation placements. This generates a database of results for specific combinations which are then regressed using algebraic or statistical methods.

The resulting regressed forms are then implemented into BPS tools which can use them to rapidly calculate $q_{e \to grd}$ using a simple function that commonly takes the form of:

$$q_{e \to grd} = A + B \cdot \sin(\omega t - C) \tag{12.2}$$

The functional form of Equation 12.2 is predicated on the approximation that the temperature of the ground surface (see Figure 12.1) varies sinusoidally over the year, and that $q_{e \to grd}$ is only strongly dependent upon this long-term variation. t is time and ω is an angular velocity of 2π rad/year. As such, this method ignores the impact of higher-frequency fluctuations in the ground-surface temperature caused by factors such as solar radiation, convection, and longwave radiation.

A, B, and C are the functions that have been regressed from the results of the detailed numerical simulations. These can be evaluated, for example, from user-supplied data for soil properties, geometry, insulation, etc. One example of such an approach that is based upon neural-network regressions is provided by Ren *et al.* (2020).

This method is computationally efficient because it only needs to evaluate the A, B, and C functions (perhaps only once prior to the timestep simulation) and Equation 12.2. However, this efficiency comes at a cost. The method is limited to the range of parameters that were used to generate the regressions. As well, it cannot consider the impact of transient indoor conditions because the regressions are developed for steady-state.

There are also analytical approaches that have been developed to estimate $q_{e \to grd}$ using expressions like Equation 12.2. However, unlike the regression-based methods that are derived from a large number of detailed numerical simulations, these are based upon analytic solutions for simplified—and sometimes unrealistic—situations.

You will discover whether your chosen BPS tool supports a regression or analytical approach during one of this chapter's simulation exercises.

12.5 SIMPLIFIED APPROXIMATIONS

Many simplified techniques are used in the BPS field to approximate $q_{e \to grd}$. Sometimes users are forced to make use of these simplifications when their chosen BPS tool does not support ground models like those described in the preceding sections. But these approaches are also employed by users in certain circumstances even when their BPS tools provide more rigorous options.

The simplest—and least accurate—option is to ignore heat transfer between the building and the surrounding ground by imposing an adiabatic boundary condition at the external surface of the envelope component in contact with the ground. This can lead to significant errors when dealing with small buildings, such as that pictured in Figure 1.1, for which up to a third of the area of all external surfaces exchanging energy with the exterior environment are in contact with the ground. In such cases ignoring $q_{e \to grd}$ would be inappropriate. However, this method might be acceptable when dealing with high-rise buildings in which $q_{e \to grd}$ would represent only a small fraction of the total heat transfer to the exterior environment.

Another simplification is to treat below-grade surfaces as being exposed to the outdoor air. This is the approach you were advised to follow when you set up your *Base Case*. With this approach, the BPS tool will calculate solar absorption, convection to the outdoor air, and longwave radiation exchange with the exterior environment using the methods described in Chapters 9, 10, and 11. This introduces many obvious approximations and errors into the analysis.

A third common simplification is for users to impose a temperature on the external below-grade surface. This is illustrated in Figure 12.3 where the temperature T_{soil} is imposed at the exterior of the insulation making up the outside layer of the floor slab construction. T_{soil} may be time-invariant or values may be prescribed for each month of the year. A common approach is to vary T_{soil} as a function of the indoor temperature. With this method heat transfer in the soil is not predicted with a model, but rather it relies upon the user

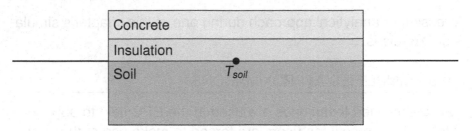

Figure 12.3: Imposing the soil temperature as a boundary condition

to prescribe the impact of the surrounding soil through the choice of appropriate values for T_{soil}. This is highly problematic because many factors determine this temperature, such as conditions inside the building, the floor construction, the soil composition, weather conditions, and the building site.

You will experiment with some of these simplified approximations during this chapter's simulation exercises.

12.6   REQUIRED READING

Reading 12–A

Chen (2015) categorizes the methods that have been developed for calculating heat transfer between building surfaces and the surrounding ground, before presenting an analytical solution for steady-state heat transfer between a slab-on-grade building and the ground.

Read Section 1 of this article and find answers to the following questions:

1. What are the five groupings the author uses to categorize the methods for calculating heat transfer between building surfaces and the surrounding ground? Compare these to the methods described in Sections 12.2 through 12.5.

2. What are the strengths and limitations of each grouping?

3. When would it be appropriate to apply a regression-based approach? What are the principal limitations of this approach?

4. What effects are not adequately considered with one-dimensional dynamic approaches?

5. What limits wider use of detailed numerical methods?

Reading 12–B

Kruis and Krarti (2017) treat methods for estimating heat transfer to the ground using two-dimensional numerical calculations. They describe some of the methods that had been previously developed and propose a new technique.

Read Sections 1, 2, 5, and 6 of this article in detail, and briefly review Sections 3 and 4. Find answers to the following questions:

1. Why is it so important in BPS to have methods to approximate three-dimensional effects using two-dimensional calculations?

2. Describe some of the availble two-dimensional approximation methods.

3. Which two-dimensional approximation methods compared favourably with the three-dimensional results?

4. The new *boundary layer adjustment method* proposed by Kruis and Krarti was found to have a lower mean bias deviation from the three-dimensional results than the existing two-dimensional approximation methods. Why?

5. In terms of computer time, how much faster are two-dimensional calculations of ground heat transfer compared to three-dimensional calculations?

12.7 SOURCES FOR FURTHER LEARNING

- Ren *et al.* (2020) present a correlation-based method for predicting heat transfer to the ground from slab-on-grade and basement foundations. These correlations are based upon detailed three-dimensional numerical simulations.

- Kruis and Krarti (2015) describe a finite-difference numerical model for calculating heat transfer between building surfaces and the ground. They compare and contrast various numerical solution schemes in terms of accuracy, computational requirements, and stability, and also examine initialization methods.

12.8   SIMULATION EXERCISES

Revert your BPS tool inputs to represent once again the *Base Case* described in Section 1.9, including any refinements or corrections you made following the previous exercises. Perform an annual simulation to produce a fresh set of *Base Case* results for use in the following exercises.

Exercise 12–A

In the *Base Case* the underside of the floor was exposed to the outdoor air. This is one of the simplified approximations described in Section 12.5 that is commonly applied by BPS users. Now apply the simplest option described in that section by making the floor adiabatic, and perform another annual simulation using the same timestep.

What impact does this change have upon the annual space heating load? And upon the annual space cooling load? Are these results in line with your expectations? Comment on the appropriateness of applying the adiabatic approximation in situations such as this.

Exercise 12–B

Now configure your tool to apply the third simplified approximation described in Section 12.5. Prescribe a constant value of 6.8 °C for T_{soil} (see Figure 12.3) and perform another annual simulation using the same timestep. This is equal to the annually averaged outdoor air temperature in the Ottawa CWEC-2016 weather file.

What impact does this change have upon the annual space heating load? And upon the annual space cooling load?

Now conduct another simulation with a T_{soil} that is 2 °C warmer. What impact does this change have upon the annual space heating load? And upon the annual space cooling load?

Comment on the sensitivity of simulation predictions to the user's choice of T_{soil} when applying this simplified approach. What sources of information could be used to establish appropriate T_{soil} values?

Exercise 12–C

Now configure your tool to use the method described in Section 12.3. Represent the ground underneath the floor construction as a 3 m thick layer of soil with properties k =1.2 W/m K, ρ =1500 kg/m^3, and c_P =1800 J/kg K. Prescribe a constant value of 6.8 °C at the *deep ground* boundary (see Figure 12.2) and perform another annual simulation using the same timestep.

Compare the simulation predictions to those from Exercise 12–B. What impact does adding the soil have upon the annual space heating load? And upon the annual space cooling load?

Now extend the soil layer from 3 m to 4 m and perform another annual simulation using the same timestep. What impact does this have upon the annual space heating load? And upon the annual space cooling load?

Exercise 12–D

Does your BPS tool support a detailed numerical modelling method like those described in Section 12.2? Does it support a regression or analytical approach like those described in Section 12.4? Describe the available modelling options.

Consult your BPS tool's help file or technical documentation to determine how to use these facilities. Then apply the most detailed method, taking appropriate assumptions regarding the required inputs.

Perform another annual simulation using the same timestep. What impact does this change have upon the annual space heating load? And upon the annual space cooling load?

Which of your assumptions are likely to have the greatest impact upon simulation predictions? Think of some situations where the additional effort and computational time required by this more detailed modelling approach would be justified.

12.9 CLOSING REMARKS

This chapter reviewed the diverse options that are available for considering the thermal interaction between buildings and the surrounding ground. Despite the significance of this energy transfer path, many BPS tools and users do not give ground models the attention they warrant.

The techniques employed by detailed numerical methods were outlined. This is the most rigorous and accurate (if correctly applied) option, although such models are not available in many BPS tools. Less computationally demanding methods were also outlined, and you explored the use of one or more of these during the simulation exercises. You also applied some of the simplified approximations that are common in the BPS field, and realized the implications of such simplicity.

This completes our treatment of heat transfer processes occurring between the building's external surfaces and the exterior environment. Our attention now turns to heat and mass transfer occurring through the building envelope.

IV

Building envelope

CHAPTER 13

Heat transfer in opaque assemblies

THIS chapter describes the main classes of methods that are commonly employed in the BPS field for treating heat transfer and storage in opaque building envelope assemblies, such as walls and ceilings. Although the mathematical developments are quite lengthy, they are necessary to understand the limitations and strengths of each approach.

Chapter learning objectives

1. Understand the mathematical basis of response function and transfer function methods.
2. Learn how numerical methods can discretize envelope assemblies in space and time to predict their performance.
3. Appreciate the basis of lumped parameter approaches.
4. Learn how to configure BPS tools to treat thermal bridges and temperature-dependent thermophysical properties.
5. Realize the sensitivity of simulation predictions to thermophysical properties.

13.1 HEAT TRANSFER PROCESSES

Terms representing heat transfer through the opaque envelope assembly appear in both the internal surface energy balance (Equation 2.14) and the external surface energy balance (Equation 8.1).

$q_{cond,i \to m}$ is the rate of heat transfer at the current timestep from surface i to the mass of the envelope construction. This is illustrated in Figure 13.1. Depending upon the prevailing boundary conditions, this energy could be transferred to surface e, or it could be stored within the mass of the envelope assembly and later transferred to either surface e or back to surface i. The same holds for $q_{cond,m \to e}$. Only under steady-state conditions—which rarely occur due to fluctuating weather, and to a lesser extent, indoor states—will $q_{cond,i \to m}$ and $q_{cond,m \to e}$ be equal at any given timestep.

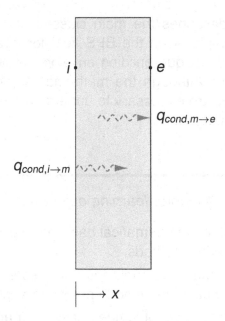

Figure 13.1: Heat transfer within an opaque envelope assembly

Heat is transferred through opaque envelope assemblies by conduction, convection, and radiation. Conduction occurs through solid structural materials, through the fibres of porous fibrous insulations, and through the solid matrix of porous foam insulations. Convection occurs through the air or gases from blowing agents trapped within foam insulations, and through the air in fibrous in-sulations. And there is longwave radiation exchange across the gaseous cells of expanded and extruded foam insulations and between solid strands of fibrous insulations. Heat transfer will be multi-directional, especially in regions where materials with vastly different thermophysical properties intersect.

Contact resistances between material layers may also exist. Ad-ditionally, convection and longwave radiation occur across spacings between material layers, such as between the OSB and cedar lay-ers of the *Base Case* wall construction (see Table 1.3).

13.2   MODELLING APPROACH

Despite this complexity, it is common practice in the BPS field to model heat transfer through opaque envelope assemblies as a conduction-only problem. With this, materials are represented with *effective* thermal conductivities that approximately represent all of the heat transfer processes occurring in reality.

Effective thermal conductivity data are available for most con-struction materials. However, the standard tests that are used to de-termine these data are typically performed at $\sim$24 °C. It has been demonstrated that some insulation materials perform very differ-ently at higher and lower temperatures because this influences the conduction, convection, and radiation processes occurring in reality (e.g. Berardi and Naldi, 2017; Berardi, 2019). Moreover, the con-densation of blowing agents at lower temperatures, freezing con-ditions, ageing, and moisture can also influence the effective con-ductivity.

In addition to treating this multi-mode heat transfer problem as a conduction-only process, methods used by BPS tools typically treat the problem as one-dimensional. Heat is assumed to flow only along the x-axis of the cross-section illustrated in Figure 13.1. In some cases the one-dimensional solution is adjusted to account for multidimensional effects such as thermal bridges.

With these simplifications, a first law energy balance can be formed that applies to any location within the envelope assembly. This takes the form of the heat diffusion equation in one dimension:

$$(\rho c_P) \cdot \frac{\partial T}{\partial t} = \frac{\partial}{\partial x} \cdot \left(k \frac{\partial T}{\partial x} \right) \tag{13.1}$$

where c_P is the specific heat (J/kg K) of the material at the point under consideration, while ρ is the density (kg/m^3) and k is the effective thermal conductivity (W/m K).

Once this governing equation is solved, the heat transfer terms required by the internal surface and external surface energy balances can be determined from the resulting temperature field for the current timestep:

$$q_{cond,i \rightarrow m} = -k_i \cdot A_i \cdot \left. \frac{\partial T}{\partial x} \right|_i \tag{13.2}$$

$$q_{cond,m \rightarrow e} = -k_e \cdot A_e \cdot \left. \frac{\partial T}{\partial x} \right|_e \tag{13.3}$$

Envelope assemblies are invariably composed of multiple materials, but each is considered to be uniform over the directions perpendicular to the x-axis. This is illustrated in Figure 13.2, which shows an envelope assembly composed of three material layers. So, when applying Equation 13.1 to the middle layer of the assembly shown in the figure, the properties of material B are used to evaluate c_P, ρ, and k. In general, k will be a function of temperature since it represents an effective thermal conductivity that accounts for all modes of heat transfer, as explained in Section 13.1.

Equation 13.1 can be solved analytically, but only for particular combinations of initial and boundary conditions. For example, if the entire assembly is initially at a uniform temperature and is then suddenly subjected to the same convection boundary condition at surfaces i and e, a solution can be determined using a mathematical technique called separation of variables. Analytical solutions can also be found for other situations, such as a step change in temperature at one boundary, or the application of a constant heat flux at a boundary.

The previous chapters have described the conditions occurring at the internal and external surfaces of the envelope assembly, including solar radiation, convection, and longwave radiation. These

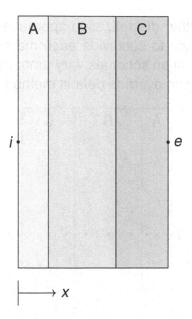

Figure 13.2: Envelope assembly composed of three material layers

create non-linear and time-varying boundary conditions at surfaces
i and *e*, and thus represent scenarios that are far more complex
than those for which analytical solutions can be found.

Therefore, a number of alternatives to analytical solutions have
been developed and applied by BPS tools to solve Equation 13.1.
These can be broadly classified into three categories. One category
applies numerical methods to discretize and solve the governing differential equation. Another class is based on semi-analytical methods that employ response functions or transfer functions. And a
third class is based upon lumped parameter methods. Each of
these categories of methods is described in the following sections.

13.3 NUMERICAL METHODS

Although different numerical discretization approaches exist, the finite difference method is most commonly employed by BPS tools.
With this, the envelope assembly is first discretized spatially into
a finite number of slices along the x-direction. This is illustrated
in Figure 13.3, where each material layer (A, B, and C) is divided

into three slices. Other discretization approaches are possible, but it is common practice to subdivide each material layer into just a few slices. Discretization schemes vary amongst BPS tools, and in some cases users can override default methods.

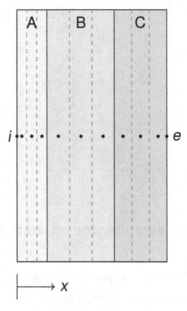

Figure 13.3: Finite difference discretization of an envelope assembly (discretized slices indicated by - - - - - and nodes by •)

A finite difference node is located at the centre of each slice[i]. The governing equation is then discretized and solved to determine the temperature at each node. The temperature throughout the envelope assembly is then approximated by a piecewise linear curve between the nodes.

The discretization of the governing equation (Equation 13.1) is demonstrated by focusing on material layer A of Figure 13.3. A larger view of this material layer is shown in Figure 13.4. We will focus on node P, which is located in the middle of the material layer. Its neighbouring nodes are indicated by E and W. The figure also illustrates the internal surface node i. The labels *east* and *west* are used to indicate the boundaries between the slices.

---

[i]Other nodal placements are possible.

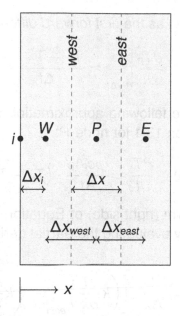

Figure 13.4: A finite difference node (P) and its neighbours (E and W)

The terms of Equation 13.1 are discretized in time and space to develop an algebraic approximation of this energy balance. Firstly, Taylor's theorem is used to approximate the temperature of node P as an infinite polynomial series of the known solution at the previous timestep $(t - \Delta t)$[ii]:

$$T_P = T_P^{t-\Delta t} + \Delta t \cdot \left.\frac{\partial T_P}{\partial t}\right|_{t-\Delta t} + \frac{(\Delta t)^2}{2} \cdot \left.\frac{\partial^2 T_P}{\partial t^2}\right|_{t-\Delta t}$$
$$+ \frac{(\Delta t)^3}{6} \cdot \left.\frac{\partial^3 T_P}{\partial t^3}\right|_{t-\Delta t} + \cdots \qquad (13.4)$$

where Δt is the duration of the simulation timestep.

The time derivative of the temperature can be approximated from Equation 13.4. One possible approach that is commonly employed is to truncate the second and higher order derivatives, an

---

[ii]When a variable is indicated without a temporal superscript, it is implied to be the value at the current simulation timestep, that is the value whose solution is sought. Therefore $T_P = T_P^t$.

approximation known as the *first forward difference* approximation:

$$\frac{\partial T_P}{\partial t}\bigg|_{t-\Delta t} \approx \frac{T_P - T_P^{t-\Delta t}}{\Delta t} \tag{13.5}$$

This leads to the following approximation of the transient term (left side) of Equation 13.1 for node P:

$$(\rho c_P)_P \cdot \frac{\partial T}{\partial t} \approx \frac{(\rho c_P)_P}{\Delta t} \cdot \left(T_P - T_P^{t-\Delta t}\right) \tag{13.6}$$

The diffusion term (right side) of Equation 13.1 can be approximated at node P by evaluating the partial derivative over the width of the slice:

$$\frac{\partial}{\partial x} \cdot \left(k\frac{\partial T}{\partial x}\right) \approx \frac{1}{\Delta x} \cdot \left[\left(k\frac{\partial T}{\partial x}\right)_{east} - \left(k\frac{\partial T}{\partial x}\right)_{west}\right] \tag{13.7}$$

where Δx is the width of the slice represented by node P and $_{east}$ and $_{west}$ indicate that the function $(k\frac{\partial T}{\partial x})$ has been evaluated at the boundaries of the slice (see Figure 13.4).

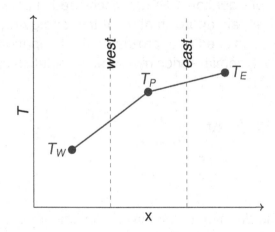

Figure 13.5: Piecewise linear approximation

The temperature gradients of Equation 13.7 can be approximated by assuming that the temperature follows a piecewise linear curve between node P and its neighbours, as illustrated in Figure

13.5. Since the thermal conductivity may vary spatially (due to temperature dependency, or different materials), some weighted averaging scheme is used to represent this property between the two nodes. For example, equally weighting[iii] the conductivity values over the two slices leads to:

$$\left(k\frac{\partial T}{\partial x}\right)_{east} \approx \left(\frac{k_P + k_E}{2}\right) \cdot \left(\frac{T_E - T_P}{\Delta x_{east}}\right) \tag{13.8}$$

where Δx_{east} is the distance between nodes P and E, as indicated in Figure 13.4.

Equation 13.8 (and a similar expression for the *west* face) can be evaluated using the known temperatures for the previous timestep. By substituting these and Equation 13.6 into Equation 13.1, an approximation of the heat diffusion equation for node P is derived:

$$\frac{(\rho c_P)_P}{\Delta t} \cdot (T_P - T_P^{t-\Delta t}) = \left(\frac{k_P + k_E}{2 \cdot \Delta x \cdot \Delta x_{east}}\right) \cdot (T_E^{t-\Delta t} - T_P^{t-\Delta t})$$
$$- \left(\frac{k_P + k_W}{2 \cdot \Delta x \cdot \Delta x_{west}}\right) \cdot (T_P^{t-\Delta t} - T_W^{t-\Delta t}) \tag{13.9}$$

Alternatively, Equation 13.8 could be evaluated using unknown temperatures at the current timestep. This would lead to a different approximation of the heat diffusion equation for node P:

$$\frac{(\rho c_P)_P}{\Delta t} \cdot (T_P - T_P^{t-\Delta t}) = \left(\frac{k_P + k_E}{2 \cdot \Delta x \cdot \Delta x_{east}}\right) \cdot (T_E - T_P)$$
$$- \left(\frac{k_P + k_W}{2 \cdot \Delta x \cdot \Delta x_{west}}\right) \cdot (T_P - T_W) \tag{13.10}$$

Equation 13.9 is known as a *fully explicit* formulation because the solution of the unknown temperature of node P for the current timestep (T_P) depends only upon the known temperature of node P and its neighbours E and W from the previous timestep's solution. This formulation is accurate, but may be unstable if too large a timestep is used.

[iii] More complex schemes that weight by mass are usually employed, but this is not shown here for the sake of clarity.

Equation 13.10 is known as a *fully implicit* formulation because the solution of T_P depends upon the yet-to-be solved temperatures of the neighbouring nodes for the same timestep. This formulation is unconditionally stable but unrealistic solutions may result with larger timesteps.

BPS tools can use either the fully explicit or fully implicit formulation, or some weighted combination of the two. A common approach is to use an equal weighting of the two schemes to achieve both stability and accuracy. Known as the Crank-Nicolson scheme, this is achieved by adding Equations 13.9 and 13.10. After some rearranging this leads to:

$$
\left[\frac{2 \cdot (\rho c_P)_P}{\Delta t} + \frac{k_P + k_E}{2 \cdot \Delta x \cdot \Delta x_{east}} + \frac{k_P + k_W}{2 \cdot \Delta x \cdot \Delta x_{west}} \right] \cdot T_P
$$
$$
- \left[\frac{k_P + k_E}{2 \cdot \Delta x \cdot \Delta x_{east}} \right] \cdot T_E - \left[\frac{k_P + k_W}{2 \cdot \Delta x \cdot \Delta x_{west}} \right] \cdot T_W
$$
$$
= \left[\frac{2 \cdot (\rho c_P)_P}{\Delta t} - \frac{k_P + k_E}{2 \cdot \Delta x \cdot \Delta x_{east}} - \frac{k_P + k_W}{2 \cdot \Delta x \cdot \Delta x_{west}} \right] \cdot T_P^{t-\Delta t}
$$
$$
+ \left[\frac{k_P + k_E}{2 \cdot \Delta x \cdot \Delta x_{east}} \right] \cdot T_E^{t-\Delta t} + \left[\frac{k_P + k_W}{2 \cdot \Delta x \cdot \Delta x_{west}} \right] \cdot T_W^{t-\Delta t}
$$

(13.11)

The left side of this equation relates the temperature of the node under consideration to the temperatures of its neighbouring nodes, all at the yet-to-be-solved current timestep. The right side of the equation contains known quantities from the previous timestep. Once an instance of this equation is formed for each of the nodes shown in Figure 13.3, the system of equations can be solved to yield the nodal temperatures for the timestep under consideration.

The solvers of some BPS tools prescribe T_i and T_e values and then solve the set of equations to determine the temperatures of the nodes located within the envelope assembly. The $q_{cond,i \rightarrow m}$ and $q_{cond,m \rightarrow e}$ heat flows are then calculated from the solved nodal temperatures using an approximation of Fourier's law:

$$
q_{cond,i \rightarrow m} = -A_i \cdot k_W \cdot \left(\frac{\partial T}{\partial x} \right)_i
$$
$$
\approx A_i \cdot k_W \cdot \left(\frac{T_i - T_W}{\Delta x_i} \right)
$$

(13.12)

where Δx_i is the distance between nodes i and W, as indicated in Figure 13.4.

The $q_{cond,i \to m}$ and $q_{cond,m \to e}$ heat flows are then communicated to the internal surface energy balance (Equation 2.14) and the external surface energy balance (Equation 8.1), where they are used to update these energy balances for the current or next timestep. (Section 2.8 discusses some of the solution possibilities.)

Other BPS tools will concurrently solve the set of heat diffusion equations with those representing the internal surface and external surface energy balances. (Again, Section 2.8 discusses some of these strategies.)

In either case, because the nodal energy balances must be re-formed and resolved at each timestep of the simulation, these numerical methods are considered to be computationally intensive. However, a benefit of the approach is that temperatures at each node are determined at each timestep of the simulation. This information may be useful, for example, for assessing condensation risk or for comparing to measurements.

Another important benefit is that thermophysical properties of the material layers can be made to vary over the course of the simulation. This would be important in the case where the effective thermal conductivity represented by k in Equation 13.11 could be represented as a function of temperature. You will explore this during one of this chapter's simulation exercises.

Although with numerical methods there is potential for spatial and temporal discretization errors, generally speaking this is rarely an issue with most constructions and with timesteps typically employed with BPS.

13.4 RESPONSE FUNCTION METHODS

In contrast to numerical methods—which solve the nodal temperatures and therefore inter-nodal heat flows at each timestep of the simulation—the response function method does not attempt to resolve the thermal state within the envelope assembly. Rather, it calculates the $q_{cond,i \to m}$ and $q_{cond,m \to e}$ as a function of current and past T_i and T_e temperatures.

Stephenson and Mitalas (1967) adapted techniques from electrical engineering and other fields to develop the method described here for calculating transient conduction through envelope assemblies. They were motivated by computational efficiency because the numerical methods described in the previous section were excessively demanding for computers at that time.

Laplace transform

The response function method is based upon a solution by Laplace transforms, in which a governing equation in the time domain is transformed to a subsidiary equation in the Laplace domain. This transformation is accomplished through an integration that is defined by:

$$f(p) = \mathcal{L}\left\{F(t)\right\} = \int_0^\infty e^{-pt} \cdot F(t)dt \qquad (13.13)$$

The symbol $\mathcal{L}$ is called the Laplace transform operator, which in the equation above is operating on the function $F(t)$ that is in the time domain. p is a complex number and $f(p)$ is the subsidiary equation in the Laplace domain.

For certain problems, the Laplace transformation results in a subsidiary equation that has a simpler functional form than the original equation in the time domain. For example, a differential equation may be simplified to an algebraic problem. Furthermore, initial conditions can be automatically incorporated into the algebraic problem.

In such cases, the resulting subsidiary equation can then be solved using algebraic manipulations. Finally an inverse transform is performed on the solution in the Laplace domain to obtain the solution in the time domain:

$$F(t) = \mathcal{L}^{-1}\left\{f(p)\right\} = \frac{1}{2\pi i} \int_{\gamma-i\infty}^{\gamma+i\infty} e^{pt}f(p)dp \qquad (13.14)$$

The symbol $\mathcal{L}^{-1}$ is called the inverse Laplace transform operator. γ is a real number and i is an imaginary number.

Laplace and inverse Laplace transforms for many common functions have been tabulated to eliminate the need to conduct the integrations given by Equations 13.13 and 13.14. For example, the following are common transforms that can be easily found in any textbook on differential equations:

$$\mathcal{L}\left\{5e^{2t}\right\} = \left(\frac{5}{p-2}\right) \tag{13.15}$$

$$\mathcal{L}^{-1}\left\{\frac{2}{p^2+4}\right\} = \sin(2t) \tag{13.16}$$

Transformation of heat diffusion equation

The heat diffusion equation that governs heat transfer within opaque envelope assemblies (subject to the simplifications outlined in Section 13.2) can be transformed to the Laplace domain by making use of tabulated transforms.

Consider the wall composed of a single material that is illustrated in Figure 13.6. By making use of tabulated Laplace transformations, Equation 13.1 can be expressed as follows if k is taken to be constant:

$$p \cdot \Theta(x, p) - T(x, t=0) = \frac{k}{\rho c_P} \cdot \frac{d^2\Theta(x, p)}{dx^2} \tag{13.17}$$

Θ is the temperature in the Laplace domain. It is a function of the location x within the material layer and of the complex number p. The initial condition in the time domain is encapsulated by the second term on the left side of the equation.

As can be seen, the Laplace operation has transformed a second-order *partial* differential equation (Equation 13.1) into a second-order *ordinary* differential equation (Equation 13.17).

Solution in Laplace domain

Although not trivial, a closed-form solution to Equation 13.17 can be found. According to Pipes (1957) the transformed heat flux at $x = 0$ can be calculated as a function of the transformed temperature and

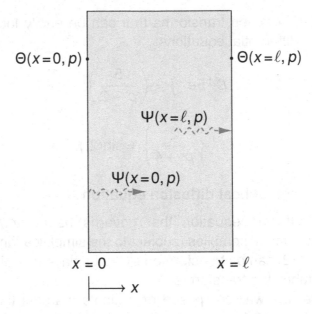

$\Theta(x=0, p)$

$\Theta(x=\ell, p)$

$\Psi(x=\ell, p)$

$\Psi(x=0, p)$

$x = 0$

$x = \ell$

x

Figure 13.6: Heat diffusion through a single material layer in Laplace domain

transformed heat flux at $x=\ell$ (refer to Figure 13.6):

$$\Psi(x=0, p) = \sqrt{k \rho c_P p} \cdot \sinh\left[\ell \cdot \sqrt{\frac{p}{k/\rho c_P}}\right] \cdot \Theta(x=\ell, p)$$
$$+ \cosh\left[\ell \cdot \sqrt{\frac{p}{k/\rho c_P}}\right] \cdot \Psi(x=\ell, p)$$

(13.18)

where Ψ is the transformed heat flux in the Laplace domain.

It can be seen from Equation 13.18 that heat transfer at the boundary ($x = 0$) can be calculated without knowledge of conditions within the material. This only requires knowledge of the conditions at the other boundary ($x = \ell$). It is important to note that this solution can only be obtained if the material properties k, ρ, and c_P are treated as constant.

This analysis can be extended to a multi-layered envelope assembly, such as that shown in Figure 13.2. After some fairly complex algebraic manipulations, expressions can be developed to calculate the transformed heat fluxes at the internal and external surfaces of the envelope assembly as a function of the transformed temperatures at these locations:

$$\Psi_i(p) = \mathcal{A}(p) \cdot \Theta_i(p) - \mathcal{B}(p) \cdot \Theta_e(p) \qquad (13.19)$$

$$\Psi_e(p) = \mathcal{B}(p) \cdot \Theta_i(p) - \mathcal{C}(p) \cdot \Theta_e(p) \qquad (13.20)$$

where $_i$ and $_e$ denote the values of the transformed quantities (Θ and Ψ) at the internal and external surfaces.

$\mathcal{A}(p)$, $\mathcal{B}(p)$, and $\mathcal{C}(p)$ are functions of the thicknesses of the assembly's material layers, as well as of their k, ρ, and c_P properties. These are complex mathematically and involve the combination of hyperbolic sine and cosine functions for each layer in the form of those appearing in Equation 13.18. But because the thermophysical properties are assumed to be constant in time, for a given wall assembly the only independent variable of the $\mathcal{A}(p)$, $\mathcal{B}(p)$, and $\mathcal{C}(p)$ functions is p.

Boundary conditions

The $\Theta_i(p)$ and $\Theta_e(p)$ terms of Equations 13.19 and 13.20 represent boundary conditions. Therefore, in order to solve these equations the surface temperatures in the time domain (T_i and T_e)[iv] must be represented and then transposed to the Laplace domain.

The common approach is to represent these continuous temperature functions as a summation of triangular pulses. Figure 13.7 illustrates how a continuous boundary condition (T_e in this example) can be represented as a summation of triangular pulses. The continuous boundary condition is shown in the left of the figure. Five triangular pulses are shown on the right. Each is centred on a timestep, and has a width of two timesteps and a height equal to the continuous function at that timestep. The summation of these five triangles is also shown on the right of the figure. It can be seen that this piecewise linear curve is a close approximation of the continuous curve on the left.

So the temperature boundary condition in the time domain can be represented as a summation of triangles. As it will be necessary

---

[iv] T_i and T_e are functions of time, but the independent variable t has been dropped for the sake of clarity, i.e. $T_i = T_i(t)$

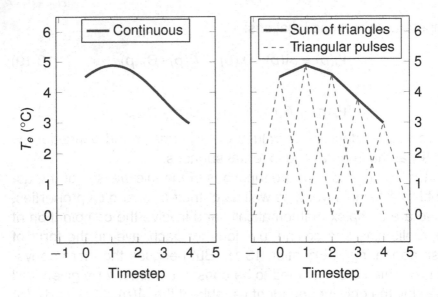

Figure 13.7: Representing a continuous temperature boundary condition (left) as a sum of triangular pulses (right)

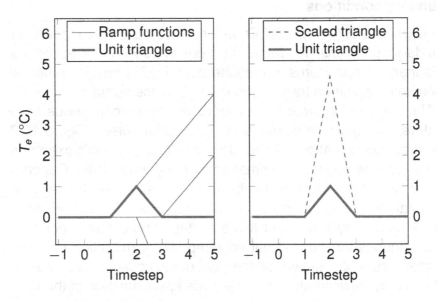

Figure 13.8: Composing a unit triangle from ramp functions (left) and linear scaling (right)

to transpose this to the Laplace domain, the next step is to represent these triangles by basic functions for which Laplace transforms have been tabulated. This is accomplished by composing a triangle of unit height and then scaling it.

This procedure is illustrated in Figure 13.8. The left of the figure shows how a triangle of unit height can be composed by summing three ramp functions. This unit triangle can then be scaled through a linear multiplier to any height. This is demonstrated in the right of Figure 13.8 where the unit triangle is scaled to represent the middle triangle of Figure 13.7.

Using this procedure, each of the triangles of Figure 13.7 can be represented by three linearly scaled ramp functions. Tabulated Laplace transforms of the ramp functions can then be summed and scaled to transpose each triangle to the Laplace domain. And summing these yields the $\Theta_i(p)$ and $\Theta_e(p)$ boundary conditions required by Equations 13.19 and 13.20.

Solution and transposition back to time domain

With the addition of the $\Theta_i(p)$ and $\Theta_e(p)$ boundary conditions, Equations 13.19 and 13.20 now express the transformed heat fluxes as complex functions of p. Recall that p is the only independent variable due to the assumption of constant thermophysical properties. The final step in the solution process is to find the inverse Laplace transforms of these equations to determine the heat fluxes at the surfaces in the time domain.

This is an involved process because tabulated inverse Laplace transforms do not exist for Equations 13.19 and 13.20 due to the convoluted form of the $\mathcal{A}$, $\mathcal{B}$, and $\mathcal{C}$ functions. Therefore, a numerical root-finding procedure is used to solve Equation 13.14. This solution is done in parts. In one part, a unit triangular pulse is imposed as the T_e boundary condition and this is represented as the summation of three ramp functions. Laplace transforms of these ramp functions are then taken to determine $\Theta_e(p)$, which is then substituted into Equation 13.19 to calculate the $\Psi_i(p)$ response. Finally, a numerical root-finding procedure is used to solve the inverse Laplace transform of $\Psi_i(p)$ using Equation 13.14 to determine the heat transfer solution in the time domain. This leads to the following partial solution:

$$q'_{cond,i\to m} = -A_i \cdot \sum_{i=0}^{\infty} Y_i \cdot T_e^{t-i\Delta t} \tag{13.21}$$

$q'_{cond,i\to m}$ is a partial solution of the heat transfer at the current timestep. As can be seen from Equation 13.21, this is given as an infinite summation of temperatures at the external surface for the current timestep ($i = 0$) as well as all preceding timesteps. The Y_i function, which applies weights to the temperature summation, is known as a *unit response function*. It is the heat transfer response to the application of a temperature boundary condition in the form of a unit triangular pulse. The elements of the Y_i function are multiplied by the magnitude of the T_e boundary condition at current and past times in order to scale the impact of a unit triangular pulse, as illustrated in Figures 13.7 and 13.8.

This Y_i response function is time-invariant and depends only upon the thickness of each of the assembly's material layers as well as their k, ρ, and c_P values. Although Equation 13.21 involves an infinite summation, in practice it can be truncated as Y_i will tend to zero when i becomes sufficiently large (the number of required terms depends upon the assembly).

Equation 13.21 is a partial solution (indicated by ') because it considers the heat transfer response at the internal surface due to the application of a unit triangle temperature boundary condition at the external surface. The impact of a temperature excitation at the internal surface must also be considered to derive the full solution. Therefore, the next step in the process is to attain another partial solution by applying a unit triangular pulse to T_i and solving the $q_{cond,i\to m}$ response. The same procedure is then applied to solve for $q_{cond,m\to e}$ to complete the set of partial solutions.

These partial solutions are then combined to yield the final solution of the heat flows in the time domain:

$$q_{cond,i\to m} = A_i \cdot \sum_{i=0}^{\infty} X_i \cdot T_i^{t-i\Delta t} - A_i \cdot \sum_{i=0}^{\infty} Y_i \cdot T_e^{t-i\Delta t} \tag{13.22}$$

$$q_{cond,m\to e} = A_e \cdot \sum_{i=0}^{\infty} Y_i \cdot T_i^{t-i\Delta t} - A_e \cdot \sum_{i=0}^{\infty} Z_i \cdot T_e^{t-i\Delta t} \tag{13.23}$$

where X_i, Y_i, and Z_i form the complete set of response functions for the envelope assembly.

As can be seen from Equations 13.22 and 13.23, the $q_{cond,i \to m}$ and $q_{cond,m \to e}$ heat flows at the current simulation timestep are calculated from the current ($i = 0$) and past ($i \geq 1$) T_i and T_e surface temperatures.

Application of these equations is computationally efficient—that was the prime motivation for the response factor method—because they are simple algebraic summations. Although the procedure for determining the response functions may be mathematically complex, X_i, Y_i, and Z_i are time-invariant and therefore need only be calculated once prior to the simulation. A cost of this computational efficiency, however, is that the temperature dependence of thermophysical properties cannot be considered.

13.5   TRANSFER FUNCTION METHODS

Not long after they proposed the response function method described in the previous section, Stephenson and Mitalas (1971) proposed an advancement—essentially a more computationally efficient arrangement—based upon Z-transforms.

Z-transforms

The method commences by approximating the Laplace transform of Equation 13.13 with an infinite summation:

$$f(p) = \int_0^\infty e^{-pt} \cdot F(t)dt$$

$$\approx \Delta t \cdot \left[F(0) + F(\Delta t)e^{-p\Delta t} + F(2\Delta t)e^{-2p\Delta t} \right. \tag{13.24}$$

$$\left. + F(3\Delta t)e^{-3p\Delta t} + \cdots + F(n\Delta t)e^{-np\Delta t} \right]$$

Equation 13.24 can be expressed more compactly by letting $z = e^{p\Delta t}$ and casting the equation as a function of the independent variable z:

$$f(p) = f(z) \approx \Delta t \cdot \left[F(0) + F(\Delta t)z^{-1} + F(2\Delta t)z^{-2} \right.$$

$$\left. + F(3\Delta t)z^{-3} + \cdots + F(n\Delta t)z^{-n} \right] \tag{13.25}$$

Equation 13.25 is known as the Z-transform of the function $F(t)$. Making use of this, the Z-transform of a function representing an envelope assembly's external surface temperature can be expressed as:

$$\Theta_e(z) = \Delta t \cdot \left[T_e(0) + T_e(\Delta t)z^{-1} + T_e(2\Delta t)z^{-2} \right.$$
$$\left. + T_e(3\Delta t)z^{-3} + \cdots + T_e(n\Delta t)z^{-n} \right] \tag{13.26}$$

And the Z-transform of a function representing the heat flow from the internal surface to the mass of the envelope assembly (see Figure 13.1) can be expressed as:

$$\Psi_i(z) = \Delta t \cdot \left[q_{cond,i \to m}(0) + q_{cond,i \to m}(\Delta t)z^{-1} \right.$$
$$+ q_{cond,i \to m}(2\Delta t)z^{-2} + q_{cond,i \to m}(3\Delta t)z^{-3} \tag{13.27}$$
$$\left. + \cdots + q_{cond,i \to m}(n\Delta t)z^{-n} \right]$$

Weighting factors

The above Z-transforms approximate the Laplace transforms that were treated in the previous section. Equation 13.26 is known as the *excitation* at the external surface, while Equation 13.27 is known as the *response* at the internal surface. The Z-transfer function of the *system* is the ratio of the two, that is, the ratio of the response to the excitation:

$$K(z) = \frac{\Psi_i}{\Theta_e} \tag{13.28}$$

A specific functional form is then imposed for the above Z-transfer function of the system:

$$K(z) = \frac{a_0 + a_1 z^{-1} + a_2 z^{-2} + a_3 z^{-3} + \cdots}{b_0 + b_1 z^{-1} + b_2 z^{-2} + b_3 z^{-3} + \cdots} \tag{13.29}$$

The a_i and b_i coefficients of this equation are known as *weighting factors*. Their values must be chosen to respect the equality between Equations 13.28 and 13.29.

Equations 13.26 and 13.27 are substituted into Equation 13.28 and the resulting $K(z)$ function replaces the left side of Equation

13.29. We now have an expression that includes a quotient of polynomials involving T_e and $q_{cond,i \to m}$ on the left and a quotient of polynomials involving a_i and b_i weighting factors on the right.

Partial solution

After considerable algebraic manipulation, a partial solution can be derived that does not require evaluation of the quantity z:

$$q'_{cond,i \to m} = \frac{A_i}{b_0} \cdot \left[\sum_{i=0}^{n} a_i \cdot T_e^{t-i\Delta t} - \sum_{i=1}^{n} b_i \cdot q_{cond,i \to m}^{t-i\Delta t} \right] \quad (13.30)$$

$q'_{cond,i \to m}$ is a partial solution of the heat transfer at the current timestep. As can be seen from Equation 13.30, this is given as a weighted summation of temperatures at the external surface for the current ($i = 0$) and past ($i \geq 1$) timesteps, as well as heat transfer values from past timesteps.

Contrast this with the similar partial solution from the response function method (Equation 13.21). The response function solution is also a weighted summation, but it contains only current and past temperatures, and not heat transfer values. In practice the a_i and b_i weighting factors diminish to negligible values at a much faster rate than do the elements of the Y_i response factors. So in practice the summations of Equation 13.30 need consider many fewer terms than the summation of Equation 13.21. This is the primary advantage of this method, which led to its wide adoption by many of the earliest BPS tools.

Transfer functions

The solution shown in Equation 13.30 is partial (indicated by ') because it considers the heat transfer response at the internal surface due to the application of a temperature excitation at the external surface. The solution must be completed by analyzing the other parts, as was done with the response function method in the previous section.

The combination of these partial solutions leads to expressions that calculate the surface heat flows at the current timestep based upon surface temperatures at current and past times, and upon surface heat flows at past times:

$$q_{cond,i \to m} = A_i \cdot \sum_{i=0}^{n} \mathcal{X}_i \cdot T_i^{t-i\Delta t} - A_i \cdot \sum_{i=0}^{n} \mathcal{Y}_i \cdot T_e^{t-i\Delta t}$$
$$+ \sum_{i=1}^{n} \mathcal{W}_i \cdot q_{cond,i \to m}^{t-i\Delta t}$$

(13.31)

$$q_{cond,m \to e} = A_e \cdot \sum_{i=0}^{n} \mathcal{Y}_i \cdot T_i^{t-i\Delta t} - A_e \cdot \sum_{i=0}^{n} \mathcal{Z}_i \cdot T_e^{t-i\Delta t}$$
$$+ \sum_{i=1}^{n} \mathcal{W}_i \cdot q_{cond,m \to e}^{t-i\Delta t}$$

(13.32)

$\mathcal{W}_i$, $\mathcal{X}_i$, $\mathcal{Y}_i$, and $\mathcal{Z}_i$ are known as *transfer functions* (often called conduction transfer functions). These are composed from the a_i and b_i weighting factors arising from the partial solutions in the form of Equation 13.30.

Determining transfer functions from response functions

Numerous methods have been devised to calculate transfer functions, although only two are commonly employed by BPS tools.

The first—which is often referred to as the *direct root finding method*—derives transfer functions from response functions. The basis of this approach can be understood by examining Equations 13.21 and 13.30. Both are partial solutions which relate the heat transfer response at the internal surface due to a temperature excitation at the external surface. By equating these two partial solutions, the a_i and b_i weighting factors can be determined from Y_i with the aid of an additional transformation and a recursive analysis. (The interested reader is referred to Underwood and Yik (2004) for details.)

Therefore, with the direct root finding method the entire response function procedure described in Section 13.4 is first applied to determine the X_i, Y_i, and Z_i response functions. The a_i and b_i weighting factors are derived from these, and then combined to determine the $\mathcal{W}_i$, $\mathcal{X}_i$, $\mathcal{Y}_i$, and $\mathcal{Z}_i$ transfer functions.

An important limitation of this approach is that significant truncation and round-off errors can occur in the case of thermally massive envelope assemblies, particularly when short timesteps are employed. To mitigate this problem, some BPS tools using this approach will restrict simulations to relatively coarse timesteps (perhaps 30 minutes or longer). As such coarse timesteps can impose significant limitations, some tools allow the use of a short timestep for simulating most of the heat and mass transfer processes, while calculating heat transfer through the opaque assemblies with a coarser timestep.

The state-space method for determining transfer functions

The second method commonly used in BPS for determining transfer functions is known as the *state-space* method. Based on the work of Ceylan and Myers (1980), the method commences with discretizing the envelope assembly using a finite difference method much like the one described in Section 13.3.

An algebraic approximation of the governing heat diffusion equation (Equation 13.1) is established for each node. However, rather than approximating the transient term using discretization techniques (see Equation 13.6), the time derivatives are retained within the resulting matrix of equations.

Unlike numerical methods, the matrix of equations is not solved to predict the nodal temperatures located within the envelope assembly. Rather, through the use of matrix algebra these nodal temperatures are absorbed and the set of equations is rearranged to yield a solution for $q_{cond,i \to m}$ and $q_{cond,m \to e}$ in the form of Equations 13.31 and 13.32.

The $\mathcal{W}_i$, $\mathcal{X}_i$, $\mathcal{Y}_i$, and $\mathcal{Z}_i$ transfer functions are then determined directly from this numerical solution, thus eliminating the need to calculate the X_i, Y_i, and Z_i response functions. As shown by Delcroix *et al.* (2013), the state-space method produces similar results to the direct root finding method but allows simulations to be conducted with somewhat shorter timesteps. You will learn more about this through one of this chapter's required readings.

Since the numerical calculations need to be performed only once prior to the timestep simulation to establish the transfer functions, the state-space method is computationally more efficient than numerical methods. However, as with the direct root finding method (and other transfer function methods), it treats k, ρ, and c_P as time-invariant.

13.6 LUMPED PARAMETER METHODS

Because of its computational simplicity, the lumped parameter method is sometimes used when a large number of simulations must be conducted very rapidly, possibly with short timesteps. This might be required for the study of building control systems or for model predictive control, for example.

This treats the envelope assembly as being composed of a small number of uniform temperature *lumps*. Heat transfer and storage by the lumps is approximated using an electrical circuit analogy. This is illustrated in Figure 13.9. In this case the envelope assembly is represented with a first-order lumped parameter model wherein the entirety of the material layers is taken to be one lump.

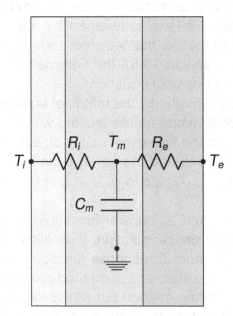

Figure 13.9: First-order lumped parameter method

T_m represents the fictitious uniform temperature of the envelope assembly, and R_i and R_e represent steady-state thermal resistances between T_m and the surface temperatures (m^2 K/W). The energy storage of the envelope assembly is considered through the capacitance parameter, C_m, which is a summation of the product of ρ, c_P, and volume of the material layers (J/K).

The overall steady-state thermal resistance of the envelope assembly is calculated and this is divided into R_i and R_e. Various techniques are available for apportioning this quantity.

Since the entire envelope assembly is treated as a single lump, the governing heat diffusion equation (Equation 13.1) reduces to a very simple form that can be rapidly calculated to determine the fictitious temperature of the assembly:

$$\frac{T_m - T_m^{t-\Delta t}}{\Delta t} = \frac{A}{C_m} \cdot \left[\frac{T_e - T_m}{R_e} - \frac{T_m - T_i}{R_i} \right] \tag{13.33}$$

The required heat transfer terms are then calculated:

$$q_{cond,i \to m} = A_i \cdot \frac{T_i - T_m}{R_i} \tag{13.34}$$

$$q_{cond,m \to e} = A_e \cdot \frac{T_m - T_e}{R_e} \tag{13.35}$$

The implications of the coarse representation of the lumps is evident. To partially mitigate this, sometimes second or higher order schemes are used wherein the envelope assembly is represented by more than one lump.

Tools that apply lumped parameter methods sacrifice a certain degree of accuracy to achieve computational efficiency. This may be a reasonable tradeoff in applications like model predictive control, where a large number of simulations over a short time horizon might be conducted to establish appropriate building control strategies. However, this technique is not commonly used by BPS tools that are used to conduct longer-term simulations.

13.7   MULTIDIMENSIONAL EFFECTS

All the methods described in this chapter for calculating transient conduction through envelope assemblies are based upon solving the one-dimensional heat diffusion equation (Equation 13.1). Heat transfer will deviate from this idealization when there are alternate flow paths with differing thermal characteristics.

Such multidimensional behaviour will be most pronounced in regions where materials with vastly different thermophysical properties intersect. This is commonly referred to as *thermal bridging*. The most significant thermal bridges tend to be caused by structural elements made from metals, concrete, and wood, which have higher thermal conductivities than surrounding insulation materials.

Very few BPS tools treat these multidimensional effects explicitly. Some allow the description of so-called *point* and *linear* anomalies that cause thermal bridging. These are essentially prescribed heat loss coefficients (W/K) that are simply multiplied by the the indoor–outdoor temperature difference, and the resulting rate of heat transfer is used to augment $q_{cond,i \to m}$ or $q_{cond,m \to e}$.

This steady-state approach relies upon the user to prescribe appropriate coefficients for each envelope assembly. Catalogues of coefficients of point and linear anomalies have been produced for various construction practices. Alternatively, the user can turn to an external detailed tool to calculate values themselves.

Many BPS tools provide no facilities whatsoever to treat thermal bridges. Rather they rely upon the user to either subdivide the envelope assembly or to modify the thermophysical properties of materials to approximately account for thermal bridges. These two approaches are illustrated in Figures 13.10 and 13.11. The actual construction of an envelope assembly under consideration is shown on the left of both figures. It can be seen that the continuity of the insulation layer (middle layer) is interrupted by structural elements, whose relatively high thermal conductivity results in thermal bridges.

The increased heat transfer caused by the thermal bridges is approximated in the modelling approach illustrated in Figure 13.10 by disregarding the structural elements and by reducing the thickness of the insulation layer (right side of figure). A different approach is illustrated in the right side of Figure 13.11. In this case, two separate envelope assemblies are used: one contains insula-

Actual   Modelled

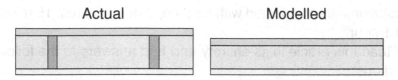

Figure 13.10: Approximating impact of thermal bridges by adjusting thermophysical properties (structural elements indicated by ■ and insulation by ▢)

Actual   Modelled

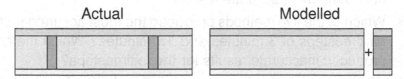

Figure 13.11: Approximating impact of thermal bridges by subdividing assembly (structural elements indicated by ■ and insulation by ▢)

tion without the structural elements and the other contains the structural elements but no insulation.

In many cases users perform calculations using either the so-called *parallel path* method or the *isothermal planes* method (ASHRAE, 2017) to determine the envelope assembly's overall heat loss coefficient under steady-state conditions (other estimation approaches are also available.) Using this result, the user can then determine appropriate dimensions and thermophysical properties for representing the envelope assembly in the BPS tool using either the approach illustrated in Figure 13.10 or that illustrated in Figure 13.11. Other approximation approaches are also in use.

13.8 REQUIRED READING

Reading 13–A

Mazzarella and Pasini (2015) examine the performance of five methods—three finite-difference schemes and two transfer function approaches—for calculating heat transfer through opaque envelope assemblies. They compare predictions from these five calculation approaches to analytical solutions for five different wall assemblies.

Simulations are conducted with timesteps of 3 minutes, 15 minutes, and 1 hour.

Read this article in its entirety and find answers to the following questions:

1. Which finite-difference discretization schemes were examined?

2. Which of the finite-difference discretization schemes produced the most accurate results?

3. Which of the five methods produced the most accurate results for timesteps of 3 minutes and 15 minutes? Which methods produced inaccurate results for these timesteps?

4. Which methods produced accurate results for timesteps of 1 hour?

5. Which of the five wall assemblies examined resulted in the greatest calculation errors? Why? Which method provided the most accurate results for this wall assembly?

Reading 13–B

Berardi and Naldi (2017) discuss the temperature dependence of the effective thermal conductivity of common insulation materials and provide measured data for four insulation materials over a wide range of temperatures. They also conduct simulations in two climate regions to assess the impact of neglecting this temperature dependency.

Read this article in its entirety and find answers to the following questions:

1. What are the four insulation materials they tested?

2. Which material was found to have a higher effective thermal conductivity when the temperature decreased below $0\,°C$?

3. Describe the wall and roof assemblies they considered in their simulations.

4. Approximately how much higher is the effective thermal conductivity of XPS insulation at $30\,°C$ compared to at $-10\,°C$?

5. Neglecting the temperature dependence of the effective thermal conductivity of the insulation materials led to some

significant errors in their simulation predictions. What range of errors did they find for the wall assemblies?

6. Why did neglecting the temperature dependency have a much greater impact on the simulation predictions of the roof assemblies?

13.9 SOURCES FOR FURTHER LEARNING

- Delcroix *et al.* (2013) compare the performance of direct root finding and state-space methods for determining transfer functions.
- Prada *et al.* (2014) examine the impact of uncertainty in thermophysical properties on the predictions from finite-difference and transfer function approaches for calculating heat transfer through opaque envelope assemblies.
- Tabares-Velasco and Griffith (2012) present a suite of analytical and numerical test cases for verifying and diagnosing models for predicting heat transfer through opaque envelope assemblies.
- Strachan *et al.* (2016) describe an empirical validation study in which 21 teams simulated the performance of two identical full-size buildings. They provide some interesting commentary about the wide divergence in approaches used by the teams for treating the thermal bridge details that were provided in the specification describing the buildings.
- Both Clarke (2001) and Underwood and Yik (2004) provide detailed descriptions of response function, transfer function, and numerical methods for calculating heat transfer through opaque envelope assemblies.

13.10 SIMULATION EXERCISES

Revert your BPS tool inputs to represent once again the *Base Case* described in Section 1.9, including any refinements or corrections you made following the previous exercises. Perform an annual simulation to produce a fresh set of *Base Case* results for use in the following exercises.

Exercise 13–A

Consult your BPS tool's help file or technical documentation to determine its default approach for modelling heat transfer through opaque envelope assemblies. Does your tool impose restrictions on the timesteps that can be used for simulations? What are the reasons for these restrictions? (Refer to the theory presented in this chapter.)

Does your tool support optional methods? For example, if your tool employs a numerical method can you control its temporal and spatial discretization schemes (refer to Section 13.3)?

If you used your BPS tool's default method in the *Base Case*, then configure it now to use an optional method. Otherwise, configure it to now use its default method. Perform another annual simulation using the same timestep.

Compare the results of the two simulations. What impact does this change have upon the annual space heating load? And upon the annual space cooling load? Was there an appreciable impact upon the time required to perform the simulations?

Exercise 13–B

Revert to your *Base Case*.

The thermophysical properties of the two material layers forming the floor construction of the *Base Case* were defined in Table 1.4. Now increase the apparent thermal conductivity of the concrete by 15 % and perform another simulation using the same timestep.

Revert to your *Base Case*, increase the concrete's density by 15 %, and perform another simulation.

Revert to your *Base Case*, increase the concrete's specific heat by 15 %, and perform another simulation.

What impact does each of these property changes have upon the annual space heating load? And upon the annual space cooling load? Are these results in line with your expectations? Think of a situation where altering the properties of a thermally massive element such as this would have a greater impact.

Exercise 13–C

Revert to your *Base Case* and then repeat the above analysis by increasing the apparent thermal conductivity, density, and specific heat of the floor's XPS insulation layer by 15%, one at a time.

What impact does each of these property changes have upon the annual space heating load? And upon the annual space cooling load?

Contrast the results from Exercise 13–B and Exercise 13–C. Which property changes had the greatest impact? Which had minimal impact? Explain these observations. What generalizations can you draw from your results?

Exercise 13–D

Revert to your *Base Case*.

The properties of the materials forming the wall assembly of the *Base Case* were defined in Table 1.3. You were told to assume that each material layer was homogeneous and to ignore the thermal impact of building structural components, fasteners, and other thermal bridges.

You will now consider the impact of thermal bridges caused by some of the wall's structure. The wall's structure is composed of a wood frame that uses studs as the vertical framing members. These studs are placed within the glass fibre layer, as illustrated in Figure 13.12. They measure 40 mm by 140 mm and are horizontally spaced at intervals of 600 mm.

Because the apparent thermal conductivity of the wood studs is higher than that of the surrounding glass fibre insulation, the studs form thermal bridges and result in multidimensional heat transfer.

What options does your chosen BPS tool offer for treating such thermal bridges? If your tool considers point or linear thermal bridge anomalies, then select appropriate inputs for this situation. You may have to consult a catalogue of coefficients or you may have to make assumptions about the thermophysical properties of the wood studs.

Alternatively, apply one of the approaches described in Section

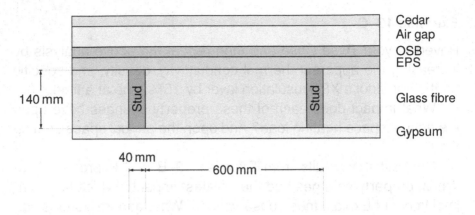

Figure 13.12: Plan view of wall assembly for Exercise 13–D

13.7 and illustrated in Figures 13.10 and 13.11. Again, make any appropriate and necessary assumptions.

Perform a simulation using the same timestep. Compare the results to those of the *Base Case*. What impact does this have upon the annual space heating load? And upon the annual space cooling load?

What other thermal bridges may exist in the *Base Case*? Discuss the significance of treating thermal bridges and the implications of ignoring them.

Exercise 13–E

Thermal conductivities were provided in Section 1.9 for all of the materials forming the *Base Case* envelope assemblies. As explained in Sections 13.1 and 13.2, these effective thermal conductivities approximately represent all of the heat transfer processes occurring in reality and are typically measured at ∼24 °C. You learned about the temperature dependence of some insulation materials through Reading 13–B.

Does your chosen BPS tool provide facilities for considering the temperature dependence of the effective thermal conductivity of materials? If not, explain why. If it does, then use the measured data provided in Table 1 of Reading 13–B to characterize the temperature dependence of the floor's XPS insulation. Conduct a simulation with the same timestep.

Compare the results to those of the *Base Case*. What impact does this have upon the annual space heating load? And upon the annual space cooling load?

Discuss the significance of ignoring the temperature dependence of the effective conductivity of insulation materials.

13.11 CLOSING REMARKS

This chapter described the main classes of methods that are commonly employed for calculating heat transfer through opaque envelope assemblies. It explained that in reality this heat transfer involves all modes (conduction, convection, and radiation) and is multidimensional, but that most BPS tools treat this as a conduction-only one-dimensional problem.

The computationally efficient response function and transfer function methods were explained. These impose an important restriction in that the temperature dependence of thermophysical properties cannot be considered. In some cases they are also limited to fairly coarse simulation timesteps. However, when thermophysical properties are only weakly dependent upon temperature, the response function and transfer function methods can achieve a high degree of accuracy.

Numerical methods were also explained, and you came to realize why these impose a greater computational burden. Although discretization errors can occur if numerical methods are not correctly configured, this approach offers some important advantages, in that it can be exercised at shorter timesteps and the temperature dependence of thermophysical properties can be considered. This can be an important consideration for some insulations, and is essential for representing phase-change materials.

Through the required readings you came to understand some of the advantages and limitations of these approaches, and developed an appreciation for the importance of considering the temperature dependence of thermophysical properties. You discovered the default and optional methods available in your chosen BPS tool through the simulation exercises, and developed techniques for dealing with thermal bridges.

CHAPTER 14

Heat transfer in transparent assemblies

THIS chapter describes the methods that are used to model heat transfer through transparent envelope assemblies (windows and skylights). These assemblies are commonly composed of insulated glazing units contained within structural framing.

Chapter learning objectives

1. Understand how the previously described internal surface and external surface energy balances can be extended to consider additional heat transfer processes occurring in transparent assemblies.
2. Comprehend the options for calculating solar transmission and absorption by glazings, and the approximations used to account for off-normal solar irradiance.
3. Grasp how heat transfer between glazings can be calculated.
4. Appreciate the techniques that can be used to account for the presence of blinds and drapes.
5. Realize the sensitivity of simulation predictions to glazing property values.

14.1   HEAT TRANSFER PROCESSES

Chapter 3 treated the methods used to determine the rate of absorption of solar energy by internal building surfaces. It explained how the $q_{solar \to i}$ quantity required by the internal surface energy balance of Equation 2.14 can be calculated once the solar irradiance transmitted through the transparent envelope assembly ($G_{solar,window}$) is known. Refer back to Figure 3.4 which illustrates the beam and diffuse components of $G_{solar,window}$.

Chapter 9 described the approaches used to calculate $G_{solar \to e}$, the solar irradiance incident upon external surfaces (including those of transparent assemblies). The beam and diffuse components of this quantity are also illustrated in Figure 3.4. This chapter connects these two topics by explaining how transparent assemblies are modelled to calculate $G_{solar,window}$ from $G_{solar \to e}$.

The insulated glazing unit of a transparent assembly is illustrated in Figure 14.1. It will transmit a portion of the solar radiation incident upon its external surface, and this will depend upon the radiative properties of the glazings and the incident angles of the beam and diffuse irradiance. Some of the incident solar irradiance will also be absorbed by the glazings. This is illustrated in Figure 14.1, where it can be seen that each of the two glazings absorb a portion of the incident solar irradiance.

The internal surface of the innermost glazing layer exchanges energy with the zone air through convection. It also exchanges energy with all other internal building surfaces through longwave radiation exchange, and possibly with internal heat sources and HVAC components that emit longwave radiation. Likewise, the external surface of the outermost glazing layer exchanges energy with the exterior environment through convection and longwave radiation.

Energy is transferred through each glazing via conduction and can be stored in the glazing's thermal mass. As shown in Figure 14.1, there will be convection heat transfer from the glazings to the fill gas contained within the insulated glazing unit. A glazing will also exchange longwave radiation with other glazings, but this will be limited to adjacent glazings since glass is opaque in the longwave spectrum.

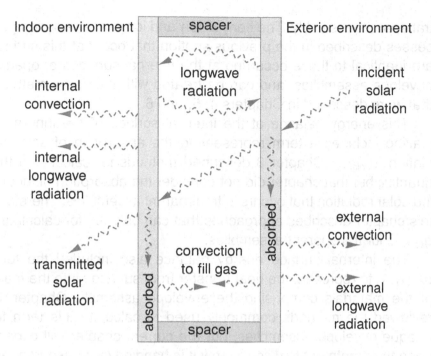

Figure 14.1: Heat transfer processes through a transparent assembly (cross-sectional view)

Energy will be transmitted across the spacers that separate the glazings, and also to the frame (not shown in Figure 14.1) that contains the insulated glazing unit. Depending upon the materials and construction of these components, this energy exchange could involve conduction, convection, and longwave radiation processes, and will be multidimensional.

The presence of blinds or drapes can influence these modes of heat transfer and add additional paths, as you will discover through this chapter's required reading.

14.2 MODELLING APPROACH

As with heat transfer through opaque envelope assemblies, the heat transfer through transparent assemblies is commonly treated as one-dimensional.

An energy balance in the form of Equation 2.14 can be established at the internal surface of the innermost glazing of the transparent assembly. The convection and longwave radiation processes described in the previous section that occur at this surface are identical to those occurring at the internal surfaces of opaque envelope assemblies, and can be treated with the same methods that were described in Chapters 4, 5, and 6.

This energy balance at the internal surface of the innermost glazing includes a term representing the absorption of solar radiation, $q_{solar \rightarrow i}$. Chapter 3 described methods for calculating this quantity, but that chapter did not consider the absorption of incoming solar radiation that occurs in transparent assemblies. Therefore, this chapter describes approaches that can be used for calculating $q_{solar \rightarrow i}$ for transparent assemblies.

The internal surface energy balance also included the term $q_{cond,i \rightarrow m}$ to represent the heat transfer from surface i into the mass of the materials comprising the envelope assembly. Chapter 13 reviewed the methods commonly used to calculate this term for opaque envelope assemblies, but the current chapter will expand upon that treatment to discuss how it is handled for transparent assemblies.

In a similar fashion, an energy balance in the form of Equation 8.1 can be established at the external surface of the outermost glazing of the transparent assembly. The convection and longwave radiation processes described in the previous section that occur at this surface are identical to those occurring at external surfaces of opaque envelope assemblies, and can be treated with the same methods that were described in Chapters 10 and 11. But as with the internal surface of the transparent assembly, alternate approaches—treated in this chapter—are required to consider the $q_{solar \rightarrow e}$ and $q_{cond,m \rightarrow e}$ terms of the external surface energy balance.

The following sections also discuss the treatment of the other heat transfer processes mentioned in the previous section. This includes solar transmission through transparent assemblies, convection between glazings and fill gas, longwave radiation exchange between glazings, and heat transfer through spacers and frames.

14.3 SOLAR TRANSMISSION AND ABSORPTION

Various methods are used to calculate the transmission of solar irradiance through transparent assemblies, and to determine the rate of energy absorption on each glazing.

Ray tracing

Many BPS tools employ a ray-tracing procedure to calculate the rate of absorption of incident solar energy by each glazing, and the amount that is transmitted through the assembly to the building interior. This requires the user to provide radiative property data in the solar spectrum for each of the assembly's glazings, such as that provided in Table 1.6 for the *Base Case*. Some tools provide databases of commercially available glazings that contain the required data to facilitate data entry.

Consider the double-glazed assembly illustrated in Figure 14.2. It is common practice to number the surfaces as in this figure, where surface 1 is the outermost surface. Recall that $G_{solar \to e}$ is the solar irradiance incident on the external surface of the assembly and that $G_{solar,window}$ is the irradiance that is transmitted through the assembly to the indoor environment (see Figure 14.2).

A ray-tracing analysis can be conducted to determine $G_{solar,window}$ from $G_{solar \to e}$. This requires knowledge of the reflectivity and absorptivity in the solar spectrum of each glazing surface, as well as the transmissivity of each glazing. Referring to the outermost glazing in Figure 14.2, $\rho^{(1)}_{solar,\theta}$ is the reflectivity of surface 1 to incoming solar irradiance at an incidence angle of θ, while $\rho^{(2)}_{solar,\theta}$ is the reflectivity of surface 2 to outgoing solar irradiance. The solar transmissivity of the glazing at incidence angle θ is given by $\tau^{(12)}_{solar,\theta}$.

Since radiation will either be transmitted, reflected, or absorbed, the solar absorptivity of each surface can be derived from these data (see Equation 3.7). For example:

$$\alpha^{(1)}_{solar,\theta} = 1 - \tau^{(12)}_{solar,\theta} - \rho^{(1)}_{solar,\theta} \tag{14.1}$$

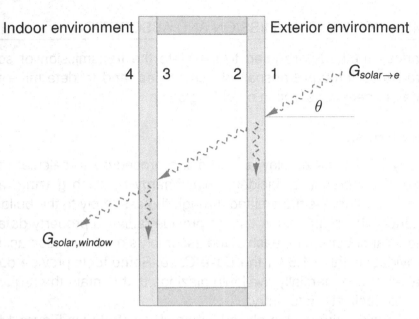

Figure 14.2: Numbering scheme used to analyze solar processes in a double-glazed transparent assembly

Transmission through assembly

A portion of the incident solar irradiance at the given incidence angle will be directly transmitted to the building interior, that is transmitted through each glazing layer without reflection. This is illustrated in the left of Figure 14.3. This directly transmitted radiation is given by:

$$G_{solar,window}^{t0} = G_{solar \to e} \cdot \tau_{solar,\theta}^{(12)} \cdot \tau_{solar,\theta}^{(34)} \qquad (14.2)$$

where ^{t0} indicates irradiance that is transmitted to the indoor environment without reflection.

Some irradiance will also be transmitted to the building interior after inter-reflections by the glazing layers. The right side of Figure 14.3 illustrates irradiance that is transmitted through the outermost glazing and that is eventually transmitted through the innermost glazing to the indoor environment after reflecting on surface 3 and then on surface 2. This component of the transmission is given by:

$$G_{solar,window}^{t1} = G_{solar \to e} \cdot \tau_{solar,\theta}^{(12)} \cdot \left[\rho_{solar,\theta}^{(3)} \cdot \rho_{solar,\theta}^{(2)} \right] \cdot \tau_{solar,\theta}^{(34)} \qquad (14.3)$$

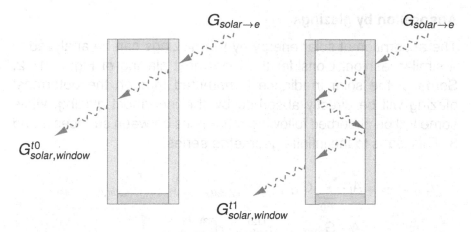

Figure 14.3: Transmission to indoor environment directly (left) and following one set of inter-reflections (right)

where ^{t1} indicates irradiance that is transmitted to the indoor environment after one set of inter-reflections.

This analysis can be continued to consider transmission occurring following two or more additional sets of inter-reflections. Summing these components leads to the total transmission of solar irradiance to the indoor environment in the form of a geometric series:

$$
\begin{aligned}
G_{solar,window} &= \sum_{i=0}^{\infty} G_{solar,window}^{ti} \\
&= \sum_{i=0}^{\infty} G_{solar \to e} \cdot \tau_{solar,\theta}^{(12)} \cdot \left[\rho_{solar,\theta}^{(3)} \cdot \rho_{solar,\theta}^{(2)} \right]^{i} \cdot \tau_{solar,\theta}^{(34)} \quad (14.4) \\
&= \frac{G_{solar \to e} \cdot \tau_{solar,\theta}^{(12)} \cdot \tau_{solar,\theta}^{(34)}}{1 - \rho_{solar,\theta}^{(2)} \cdot \rho_{solar,\theta}^{(3)}}
\end{aligned}
$$

Therefore, Equation 14.4 can be used to calculate the total transmission of solar irradiance to the indoor environment knowing $G_{solar \to e}$ and the radiative properties of each glazing layer at the prevailing angle of incidence.

Absorption by glazings

The absorption of solar energy by the glazings can be analyzed in a similar fashion. Consider the innermost glazing of Figure 14.2. Some of the solar irradiance transmitted through the outermost glazing will be directly absorbed by the innermost glazing, while some will be absorbed following reflections between surfaces 2 and 3. This leads to the infinite geometric series:

$$
\begin{aligned}
q_{solar \to 34} &= A_i \cdot \sum_{i=0}^{\infty} G_{solar \to e} \cdot \tau_{solar,\theta}^{(12)} \cdot \alpha_{solar,\theta}^{(3)} \cdot \left[\rho_{solar,\theta}^{(3)} \cdot \rho_{solar,\theta}^{(2)} \right]^i \\
&= \frac{A_i \cdot G_{solar \to e} \cdot \tau_{solar,\theta}^{(12)} \cdot \alpha_{solar,\theta}^{(3)}}{1 - \rho_{solar,\theta}^{(2)} \cdot \rho_{solar,\theta}^{(3)}}
\end{aligned}
$$

(14.5)

$q_{solar \to 34}$ represents the rate of energy absorption by the innermost glazing. The next section will discuss how $q_{solar \to i}$ is determined from this quantity.

The absorption by the outermost glazing can be determined in a similar fashion by considering direct absorption and absorption following reflections between surfaces 2 and 3:

$$
\begin{aligned}
q_{solar \to 12} &= A_e \cdot G_{solar \to e} \cdot \alpha_{solar,\theta}^{(1)} \\
&+ A_e \cdot \sum_{i=0}^{\infty} G_{solar \to e} \cdot \tau_{solar,\theta}^{(12)} \cdot \rho_{solar,\theta}^{(3)} \cdot \alpha_{solar,\theta}^{(2)} \cdot \left[\rho_{solar,\theta}^{(2)} \cdot \rho_{solar,\theta}^{(3)} \right]^i \\
&= A_e \cdot G_{solar \to e} \cdot \left(\alpha_{solar,\theta}^{(1)} + \frac{\tau_{solar,\theta}^{(12)} \cdot \rho_{solar,\theta}^{(3)} \cdot \alpha_{solar,\theta}^{(2)}}{1 - \rho_{solar,\theta}^{(2)} \cdot \rho_{solar,\theta}^{(3)}} \right)
\end{aligned}
$$

(14.6)

The above procedures can be extended to determine the transmission and absorption of an assembly composed of any number of glazings.

Off-normal incidence

Most available glazing data have been determined with spectrophotometers that can only measure at normal incidence. This is an important limitation because Equations 14.4 through 14.6 require the radiation properties at the prevailing incidence angle, and this angle will only rarely be normal. For example, the incidence angle of the beam irradiation to a south facing window will usually be greater than 40°.

To deal with this, most BPS tools that apply a ray-tracing approach determine off-normal radiative properties from the user-supplied normal-incidence data (refer to Table 1.6).

The transmissivity of uncoated glazings decreases only slightly from $0° < \theta < 40°$, and then drops more rapidly to zero at $\theta = 90°$. And the reflectivity remains roughly constant from $0° < \theta < 40°$ and then rises rapidly to 1 at $\theta = 90°$. Many BPS tools accurately calculate—for uncoated glazings—this angular dependence based upon fundamental optical relations using the method proposed by Furler (1991).

This method relates the angular property to the value at normal incidence. Figure 14.4 plots the result of these calculations for two uncoated, homogeneous glazings. Using this approach, the BPS tool can determine the $\tau_{solar,\theta}$, $\rho_{solar,\theta}$, and $\alpha_{solar,\theta}$ values required by Equations 14.4 through 14.6 for the prevailing incidence angle.

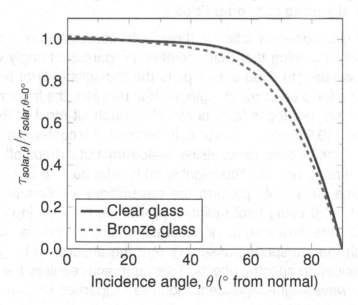

Figure 14.4: Transmissivity of two uncoated, homogeneous glazings as a function of incidence angle

However, Furler's method cannot be applied to coated glazings, which contain very thin layers that are deposited upon homogeneous glazings. (Two of the *Base Case* glazings are coated.) Since no comparable calculation approach has been devised for coated glazings, BPS tools commonly approximate the off-normal behaviour of coated glazings by assuming they behave like uncoated glazings. This is done by using a curve such as those plotted in Figure 14.4. It is common to use a relationship for bronze glass when treating a coated glazing with a low normal transmissivity, and one for clear glass for more transparent coated glazings.

Equations 14.4 through 14.6 are applied for both the beam ($G_{solar\_beam \rightarrow e}$) and diffuse ($G_{solar\_diff \rightarrow e}$) components of the incident irradiance. We previously discussed how θ is determined for the beam component (see Section 9.3). The radiative properties (such as given in Figure 14.4) are usually integrated over the hemisphere to determine appropriate values of $\tau_{solar,\theta}$, $\rho_{solar,\theta}$, and $\alpha_{solar,\theta}$ for analyzing the diffuse component.

Spectral versus total quantities

Some glazings—in particular those with coatings—are spectrally selective, meaning that their radiative properties strongly depend upon wavelength. Figure 14.5 plots the transmissivity of two glazings as a function of wavelength. As can be seen, the transmissivity of the clear glazing is fairly constant throughout most of the solar spectrum ($0.3\,\mu m \lesssim \lambda \lesssim 2.4\,\mu m$). In contrast, the coated glazing has a high transmissivity in the visible spectrum, but it drops off sharply in the infrared region. (You might want to refer back to Section 3.2.)

Some BPS tools perform the calculations of Equations 14.4 through 14.6 using total radiative properties that are input by the user. Others, however, more accurately calculate the transmission through the transparent assembly and the absorption by glazings by considering spectral effects. This approach requires the user to supply wavelength-dependent radiative properties ($\tau_{\lambda,\theta}$, $\rho_{\lambda,\theta}$, $\alpha_{\lambda,\theta}$) for each glazing[i].

---

[i]The Lawrence Berkeley National Laboratory's WINDOW tool provides a comprehensive library of glazing property data.

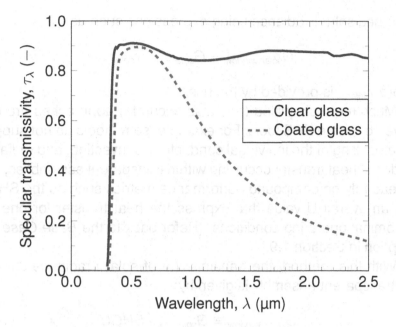

Figure 14.5: Specular transmissivity of two glazings

With the spectral approach, Equations 14.4 through 14.6 are evaluated at a number of different wavelengths using wavelength-dependent properties. A weighted integration of these results is then performed to determine the total irradiance quantities integrated over the solar spectrum (e.g. $G_{solar,window}$). You will explore this approach during one of this chapter's simulation exercises.

Other approaches

The ray-tracing approaches outlined here are not universally employed by BPS tools. In some cases the user is required to provide angular dependent solar radiation properties for the glazings and/or for the transparent assembly, such as are provided in Table 1.7. As such data are rarely available, users must usually rely upon an external tool (e.g. LBNL WINDOW) to calculate these quantities. These external tools typically employ the ray-tracing methods outlined here.

With this method, the BPS tool calculates the transmitted or absorbed solar irradiance using expressions such as:

$$G_{solar,window} = G_{solar \rightarrow e} \cdot \tau_{solar,\theta} \tag{14.7}$$

where $\tau_{solar,\theta}$ is provided by the user.

Much more simplified (and less accurate) approaches are employed by some BPS tools. For example, some tools do not support the modelling of the individual conduction, convection, and radiation modes of heat transfer occurring within transparent assemblies, and instead rely on compound performance metrics such as the SHGC and an overall U-value that express the heat transfer for one set of nominal operating conditions. (Refer back to the *Base Case* description in Section 1.9.)

With this method, the transmission of solar irradiance through the transparent assembly is given by:

$$G_{solar,window} = G_{solar \rightarrow e} \cdot SHGC \tag{14.8}$$

where the SHGC is given by the user, and may or may not vary with angle of incidence.

14.4   CONDUCTION AND STORAGE BY GLAZING

Various approaches are in use for treating energy transfer through glazings via conduction and energy storage in the glazing's thermal mass.

Some BPS tools ignore transient energy storage and treat heat transfer across the glazing as a steady-state conduction problem. Consider the innermost glazing of Figure 14.2. With the steady-state assumption, conduction across the glazing can be given by:

$$q_{cond,i \rightarrow m} = q_{cond,4 \rightarrow 3} = -q_{cond,3 \rightarrow 4} = A_i \cdot \frac{k_{34}}{\Delta x_{34}} \cdot (T_4 - T_3) \tag{14.9}$$

where k_{34} is the thermal conductivity and Δx_{34} the thickness of the glazing.

With this method the absorption of solar energy by the glazing is assumed to occur at the two surfaces of the glazing. As such, the

$q_{solar \rightarrow 34}$ calculated with Equation 14.5 is equally split between the energy balances of surfaces 3 and 4.

Other BPS tools treat each glazing layer as being at a uniform temperature, i.e. $T_4 = T_3$. All of the absorbed solar energy appears in this balance. With this approach the $q_{cond,i \rightarrow m}$ term of the internal surface energy balance is replaced by terms that consider convection to the fill gas and longwave radiation exchange with the neighbouring glazing (topics of the next section).

And some BPS tools apply the numerical or lumped parameter methods described in Chapter 13 to explicitly treat transient storage and conduction. As described above, these must also make assumptions about how to apportion the absorbed solar energy (e.g. $q_{solar \rightarrow 34}$) to the nodes or lumps.

14.5   HEAT TRANSFER BETWEEN GLAZINGS

Most BPS tools form energy balances for the glazing surfaces adjacent to the fill gas using the same type of approaches that were used to establish the internal surface energy balance of Equation 2.14. Consider surface 3 in Figure 14.2. An energy balance for this surface can be written as:

$$q_{solar \rightarrow 3} + q_{cond,m \rightarrow 3} = q_{conv,3 \rightarrow gas} + q_{lw,3 \rightarrow 2} \tag{14.10}$$

Sections 14.3 and 14.4 described approaches for determining $q_{solar \rightarrow 3}$ and $q_{cond,m \rightarrow 3}$. We will now look at how the techniques previously described in Chapters 4 and 5 can be used to establish the two terms on the right side of Equation 14.10.

A temperature difference between surfaces 2 and 3 can give rise to a buoyancy-driven recirculation within the fill gas occupying the gap between the glazings. As this fill gas is a closed system with negligible mass to store energy, the rate of convection heat transfer from surface 3 to the gas can be considered to equal that from the gas to surface 2. As such:

$$q_{conv,3 \rightarrow gas} = q_{conv,gas \rightarrow 2} = A_i \cdot h_{conv} \cdot (T_3 - T_2) \tag{14.11}$$

Determining an appropriate h_{conv} convection coefficient is key to solving Equation 14.11. This can be accomplished by returning to the methods introduced in Chapter 4. The natural convection regime can be characterized by selecting a suitable Nusselt versus Rayleigh number correlation in the form of Equation 4.7 (different tools use different correlations), and the resulting Nusselt number is used to calculate h_{conv} using Equation 4.2.

The longwave radiation exchange between surfaces 2 and 3 can be treated by returning to the methods described in Chapter 5, in particular Section 5.6. Since the height of the transparent assembly is significantly greater than the spacing between glazings and since glass is opaque to longwave radiation, the view factors (introduced in Section 5.4) between glazings can be considered to be close to unity, that is $f_{2 \rightarrow 3} = f_{3 \rightarrow 2} \approx 1$. With this, Equation 5.19, which calculates the net longwave radiation heat transfer between surfaces, can be written as:

$$q_{lw,3 \rightarrow 2} = \frac{\epsilon_{lw,2} \cdot \epsilon_{lw,3} \cdot \sigma \cdot A_i \cdot \left(T_3^4 - T_2^4\right)}{1 - (1 - \epsilon_{lw,2})(1 - \epsilon_{lw,3})} \tag{14.12}$$

The above methods are the most accurate for determining the convection and longwave radiation between glazings, but they are not universally employed by BPS tools. Rather than calculating temperature-dependent convection and longwave radiation processes using these approaches, some BPS tools use a more simplified treatment which requires the user to prescribe a time-invariant thermal resistance to represent these processes.

In this case, the combined effects of convection and longwave radiation are calculated by:

$$q_{conv,3 \rightarrow 2} + q_{lw,3 \rightarrow 2} = \frac{A_i \cdot (T_3 - T_2)}{R_{gap}} \tag{14.13}$$

where R_{gap} ($m^2\,K/W$) must be supplied by the user.

14.6 SPACERS AND FRAMES

As mentioned in Section 14.1, heat transfer through spacers and frames is multidimensional and may involve conduction, convection, and longwave radiation processes. This shares a good deal of commonality with thermal bridging through opaque envelope constructions, the subject of Section 13.7.

Indeed, the techniques used to treat thermal bridging are also often employed to account for heat transfer through transparent assembly spacers and frames. New assemblies are often added by the user to account for this heat transfer path. Tabulated data is often used to establish appropriate thermophysical properties for these assemblies. Alternatively, calculations can be performed in external tools to establish appropriate values.

During this chapter's simulation exercises you will explore the facilities available in your chosen BPS tool for treating spacers and frames.

14.7 REQUIRED READING

Reading 14–A

Lomanowski and Wright (2012) describe a methodology for modelling so-called *complex fenestration constructions*. These are transparent envelope assemblies that include shading devices, such as louvre blinds, roller blinds, or drapes.

Read the first two sections of this article and find answers to the following:

1. According to this article, BPS tools typically consider three parallel paths of heat transfer through transparent assemblies. What are these three paths?
2. Which of these three paths typically accounts for the majority of heat transfer?
3. Explain why the presence of shading devices complicates the calculation of solar transmission and absorption. Refer to Section 14.3 of this chapter in formulating your response. Which of this section's equations will be affected by the presence of blinds?
4. Explain why the presence of louvre blinds complicates the calculation of longwave radiation exchange between glazing layers. Refer to Section 14.5 of this chapter in formulating your response.
5. What is the definition of *diathermanous*?

6. Why does the methodology treated in the article employ so-called *jump resistors*?

14.8 SOURCES FOR FURTHER LEARNING

- Curcija *et al.* (2018) provide a comprehensive description of approaches for modelling heat transfer through transparent envelope assemblies.
- de Gastines *et al.* (2019) describe a common approach for modelling heat transfer through window frames and propose improvements for treating highly conductive frames, such as those made of metal.

14.9 SIMULATION EXERCISES

Revert your BPS tool inputs to represent once again the *Base Case* described in Section 1.9, including any refinements or corrections you made following the previous exercises. Perform an annual simulation to produce a fresh set of *Base Case* results for use in the following exercises.

Exercise 14–A

Section 1.9 provided data to support all of the models that are currently used by BPS tools for calculating the radiation, convection, and conduction processes that occur across transparent assemblies. From this you had to select the input data required by your chosen BPS tool and model.

Which of the provided information did you use for simulating the *Base Case*? Based on information provided in your BPS tool's help file or technical documentation, describe the model it uses to calculate heat transfer through transparent assemblies.

Does your chosen BPS tool support optional methods for modelling transparent assemblies? If so, then replace the model you employed with an alternate method provided by your tool, making use of the data provided in Section 1.9. Perform another annual simulation using the same timestep.

What impact does this choice of model have on the annual space heating load? And upon the annual space cooling load?

Exercise 14–B

Revert to your *Base Case*.

The *Base Case* window is triple-glazed and filled with argon. The inner two glazings contain low-emissivity coatings on surfaces 3 and 5, as illustrated by the dashed lines in Figure 1.3.

Now replace the argon gas fill with air and perform another simulation using the same timestep.

Revert to your *Base Case*, increase $\epsilon_{lw,front}$ for glazings 3-4 and 5-6 from 0.095 to 0.84, and perform another simulation. If your BPS tool does not use the data provided in Table 1.6 you will have to use an external tool to generate the equivalent of the data in Table 1.7 or the SHGC and U-value.

Revert to your *Base Case*, increase $\tau_{solar,\perp}$ for glazings 3-4 and 5-6 from 0.708 to 0.876, and perform another simulation. If your BPS tool does not use the data provided in Table 1.6 you will have to use an external tool to generate the equivalent of the data in Table 1.7 or the SHGC and U-value.

What impact does each of these property changes have upon the annual space heating load? And upon the annual space cooling load? Which had the greatest impact? Are these results in line with your expectations?

Exercise 14–C

Revert to your *Base Case*.

Recall that the presence of window frames and glazing spacers was ignored in the *Base Case*. Now consider that the window assembly is supported by a vinyl frame that is 50 mm wide and that has a U-value of $2\,W/m^2\,K$. The area of the glazed portion remains unchanged. Add the frame and conduct another simulation with the same timestep.

What impact does this have on the annual space heating load? And upon the annual space cooling load?

Exercise 14–D

Revert to your *Base Case*.

Now add an internal roller blind to the window which is fully deployed. The blind's solar transmittance is 0.07 and its solar reflectance is 0.61. The blind has an openness factor of 0.03 and the longwave emissivity of the blind material is 0.9.

What options does your BPS tool provide for simulating the presence of this roller blind? What additional data must you provide?

Conduct another simulation with the same timestep. What impact does the addition of the roller blind have upon the annual space heating load? And upon the annual space cooling load?

Exercise 14–E

Revert to your *Base Case*.

Section 14.3 explained that some BPS tools can consider the spectral dependency of radiative properties of glazings. The radiative properties provided in Table 1.6 were *total*, in that they were integrated over the solar spectrum.

You can only perform this exercise if your chosen BPS tool can consider spectral effects. If so, rather than using the total properties provided by Table 1.6, use spectral quantities determined with an external tool, such as LBNL WINDOW. Calculate the spectral properties assuming that glazing 1-2 is glazing ID 9801 of the International Glazing Database (IGDB), and that glazings 3-4 and 5-6 are ID 3238. These have the same total radiative properties as given in Table 1.6.

Export the spectral properties for this transparent assembly from the external tool and provide this information to your chosen BPS tool. Perform another simulation with the same timestep.

What impact does this have on the annual space heating load? And upon the annual space cooling load?

14.10   CLOSING REMARKS

This chapter described the methods that are in use for calculating heat transfer through transparent envelope assemblies. It explained how ray-tracing approaches can be employed for determining the transmission of solar radiation through such assemblies, and for determining the absorption of energy by glazings. Techniques for dealing with off-normal angles of incidence and spectral dependency were also outlined. The alternatives to these approaches that are used by some BPS tools were also described.

The variety of methods that are in use for treating other modes of heat transfer, including conduction through glazings, energy storage by glazings, longwave radiation between neighbouring glazings, and convection between glazings and fill gas were also outlined.

Through the required reading you learned about techniques that have been devised for treating shading devices, and you explored the facilities offered by your chosen BPS tool for treating these through one of the simulation exercises.

CHAPTER 15

Air infiltration and natural ventilation

THIS chapter describes methods for determining the rate of outdoor airflow through the building envelope. This includes airflow through intentional openings (natural ventilation) as well as airflow through unintended openings (infiltration).

Chapter learning objectives

1. Realize the common options for treating air infiltration and natural ventilation.
2. Understand the approaches underlying single-zone methods for calculating air infiltration.
3. Comprehend the calculation methods used by network airflow models, and realize what data they require and the effort involved in applying them.
4. Realize the impact of outdoor airflow upon simulation predictions.
5. Learn how to configure your chosen BPS tool to calculate outdoor airflow rates.

15.1   AIRFLOWS THROUGH THE BUILDING ENVELOPE

It is common to distinguish between intended and unintended air-flow through the building envelope. As explained earlier in the book (Section 2.4), airflow through intentional openings such as windows is commonly called *natural ventilation*, while airflow through unintended openings, such as cracks, holes, and imperfections in air barriers is called *infiltration*.

Since these mass flows have the same thermal impact, they are collectively represented by the $\dot{m}_{a,inf}$ term in the zone dry-air mass balance (Equation 2.3), the zone moisture mass balance (Equation 2.6), and the zone energy balance (Equation 2.13). $\dot{m}_{a,inf}$ appears within a summation in each of these balances because the zone under consideration may be receiving outdoor air through numerous openings in the building envelope.

Air will flow—from outdoors into the zone, or from the zone to outdoors—whenever a pressure difference occurs across an opening in the building envelope. The characteristics of each opening are unique and will influence these airflows. For a given pressure difference, there will obviously be less airflow through a small crack or imperfection in an air barrier than through an open window, for example.

Pressure differences across the building envelope can be induced by wind or by mechanical ventilation systems that supply or extract air from a zone. Furnaces, boilers, hot water heaters, and cooking devices that draw combustion air from the zone can also generate pressure differences across the building envelope, as can leaking ducts of air-based HVAC systems.

Hydrostatic pressure differences between indoor and exterior environments can also cause pressure differences across openings. Known as the *stack effect*, this is caused by a difference between the indoor and outdoor temperature that leads to a difference in density, and therefore a difference in the weight of a column of air on the two sides of the opening.

These pressure conditions are highly variable, depending upon the speed, direction, and turbulence of the wind. They also depend upon the building's shape, the local terrain, temperature conditions, and the functioning of combustion equipment and HVAC systems.

15.2 MODELLING OPTIONS

There are numerous options for establishing the $\dot{m}_{a,inf}$ airflow rates required by the zone mass and energy balances:

1. In some cases users ignore airflow through the building envelope, essentially setting all $\dot{m}_{a,inf}$ values to zero. This is rarely a good practice as these terms are often significant in the zone energy balance.

2. Users prescribe time-invariant values, perhaps based upon experience. (This is the approach that has been used in the simulation exercises of all the previous chapters.) This is a slight improvement upon the previous option, but such an approach can lead to considerable uncertainty because, in reality, air infiltration and natural ventilation rates can be highly variable and are often significant in the zone energy balance.

3. Users prescribe scheduled values, which perhaps vary by time of day or by season. This shares the same drawbacks as the previous approach.

4. Some BPS tools provide facilities that can be used in conjunction with one of the above options to make airflow rates vary in consequence to simulation parameters. This could be used, for example, to mimic window openings by increasing the value of $\dot{m}_{a,inf}$ when T_z rises above a user-prescribed value. However, this still relies upon the user to prescribe an appropriate value of $\dot{m}_{a,inf}$ for when the window is open, and an appropriate value for when it is closed. Its accuracy is further dependent upon specifying control setpoints that accurately reflect occupant behaviour.

5. Predicting the total air infiltration rate to the entire building ($\sum \dot{m}_{a,inf}$) with a single-zone model that responds to prevailing wind and temperature conditions.

6. Predicting airflow rates for each possible flow path (individual $\dot{m}_{a,inf}$ values) with a network airflow model.

You will explore which of these options are supported by your chosen BPS tool during this chapter's simulation exercises, and you will see first hand the impact these have upon simulation predictions.

15.3 SINGLE-ZONE MODELS

Some BPS tools include single-zone models for calculating air infiltration rates, although their applicability is usually limited to low-rise buildings. These do not calculate $\dot{m}_{a,inf}$ values for individual airflow paths, but rather determine the total air infiltration rate through the entire building envelope, $\sum \dot{m}_{a,inf}$. If the user has subdivided the building into a number of thermal zones, it is usually left to the user to configure the BPS tool to apportion $\sum \dot{m}_{a,inf}$ to each zone.

The earliest single-zone methods were based on an orifice flow assumption. Consider airflow through the opening in the building envelope that is illustrated in Figure 15.1. This mass flow is driven by the pressure difference across the opening, $\Delta P = P_e - P_i$. The local air pressure on the exterior environment side of the opening is given by P_e, while P_i represents the pressure on the indoor side, as shown in the figure. In this example, air is infiltrating from the exterior environment to the indoor environment because $P_e > P_i$.

The first law energy balance of Equation 2.7 can be applied to this situation with a number of simplifying assumptions. The process is assumed to be steady and friction is neglected. The kinetic energy of the flow entering the opening is also neglected, and it is assumed that the opening is horizontal. By recognizing that the enthalpy terms of Equation 2.7 can be expressed as $h = u + P/\rho$, where u is the internal energy (consult any thermodynamics textbook), and by neglecting the change in internal energy across the opening, the first law energy balance can be expressed as:

$$0 = \frac{P_e}{\rho_e} - \frac{P_i}{\rho_i} - \frac{V_i^2}{2} \tag{15.1}$$

By neglecting changes in density, this can be rearranged into

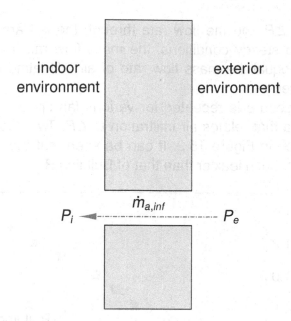

Figure 15.1: Airflow through an opening in the building envelope

the familiar form of Bernoulli's flow equation for frictionless flow:

$$\frac{P_e - P_i}{\rho} = \frac{V_i^2}{2}$$

$$\frac{\Delta P}{\rho} = \frac{V_i^2}{2}$$
(15.2)

The mass flow rate of the infiltrating air can be calculated using Equation 15.2:

$$\sum \dot{m}_{a,inf} = \rho \cdot A_{open} \cdot V_i$$

$$= A_{open} \cdot \left(2 \cdot \rho \cdot \Delta P\right)^{1/2}$$
(15.3)

Equation 15.3 is the basis of the Sherman and Grimsrud (1980) single-zone model. This model requires measurements (or estimates) of the building's airtightness determined by depressurizing[i] the building using a blower door apparatus. The apparatus' fan exhausts air from the building until steady conditions are achieved, at

[i]In some tests the building is pressurized, while in others it is depressurized.

which point ΔP and the flow rate through the fan are measured. Under these steady conditions, the mass flow rate exhausted by the fan will equal the mass flow rate of air infiltrating through the building envelope.

The procedure is repeated for various fan speeds to produce a set of data that relates air infiltration to ΔP. Two such data sets are illustrated in Figure 15.2. It can be seen that the envelope of Building A is much leakier than that of Building B.

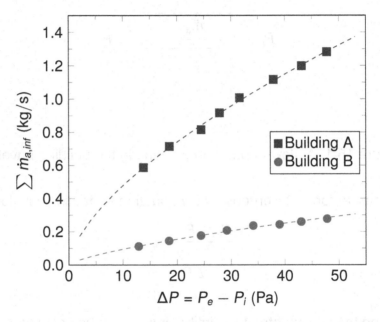

Figure 15.2: Measurements from building depressurization tests (measurements indicated by ■ and ●, the fitted power law relations by - - - - -)

For the sake of measurement accuracy these tests are usually performed at higher ΔP than experienced under normal operating conditions. The measured data are then fit to a power law relationship:

$$\sum \dot{m}_{a,inf} = \rho_{oa} \cdot C \cdot \Delta P^n \qquad (15.4)$$

C and n are empirical coefficients that are chosen to best fit the data. Figure 15.2 illustrates the fitted power law relations for these two buildings.

In the Sherman and Grimsrud method, the fitted curve of Equation 15.4 is extrapolated to estimate the air leakage at $\Delta P = 4\,\text{Pa}$:

$$\sum \dot{m}_{a,inf}\Big|_{\Delta P=4\,\text{Pa}} = \rho_{oa} \cdot C \cdot (4\,\text{Pa})^n \tag{15.5}$$

This value is then substituted into Equation 15.3, which is rearranged to determine the size of the orifice that would result in this same mass flow rate:

$$A_{open} = \frac{\sum \dot{m}_{a,inf}\Big|_{\Delta P=4\,\text{Pa}}}{\left(2 \cdot \rho \cdot 4\,\text{Pa}\right)^{1/2}} \tag{15.6}$$

A_{open} is known as the *effective leakage area*. It is the size of a frictionless orifice that produces the same air infiltration at $\Delta P = 4\,\text{Pa}$ as the sum of all the building's actual openings, and is a parameter that characterizes the building's air leakage.

During the simulation, the ΔP caused by the wind, the stack effect, and the other factors mentioned in Section 15.1 are determined, and then the air infiltration is calculated using Equation 15.3 and the user-supplied A_{open}.

Another commonly used single-zone model is that of Walker and Wilson (1998). It shares much in common with the Sherman and Grimsrud model, except that it does not use Equation 15.3, nor does it characterize the building envelope with A_{open}. Rather, it calculates the air infiltration using the power law relation of Equation 15.4. With this model, the user must supply the C and n coefficients to characterize the building.

Both the Sherman and Grimsrud and Walker and Wilson single-zone models separately calculate the air infiltration caused by the stack effect and the wind. For example, with the Walker and Wilson model:

$$\sum \dot{m}_{a,inf,stack} = \rho_{oa} \cdot f_{stack} \cdot C \cdot \Delta P_{stack}^n \tag{15.7}$$

$$\sum \dot{m}_{a,inf,wind} = \rho_{oa} \cdot f_{wind} \cdot C \cdot \Delta P_{wind}^n \tag{15.8}$$

f_{stack} is known as the *stack factor* and f_{wind} the *wind factor*. These depend upon the location (floor, ceiling, walls) of the leakage paths in the building envelope and can be calculated using relations provided by Walker and Wilson. However, this is problematic because the depressurization test described above does not identify individual leakage paths, but rather determines their collective impact. Although some BPS tools can calculate f_{stack} and f_{wind} if the user provides data on the leakage distribution, most provide default data for representative building types.

The ΔP_{stack} required by Equation 15.7 is the pressure difference caused by the stack effect acting in isolation. As Section 15.1 explained, this is caused by a difference between the indoor and outdoor temperature. In reality, the stack pressure across each opening in the building envelope will be different because it depends on the weight, and therefore the height, of columns of air on the two sides of the opening. But since single-zone models consider the collective impact of all leaks, a single value of ΔP_{stack} is determined for the entire building:

$$\Delta P_{stack} = \rho_{oa} \cdot g \cdot H \cdot \frac{|T_z - T_{oa}|}{T_z} \tag{15.9}$$

With some BPS tools the user must specify which zone's T_z to use in Equation 15.9. H is usually taken as the highest point of the building envelope's air barrier, again a quantity that the user may have to supply.

Every opening in the building envelope will, in general, experience a different pressure from the wind. This will depend upon the speed, direction, and turbulence of the wind, the building's shape, and the local terrain. But again, this being a single-zone model a single value of ΔP_{wind} is determined for the entire building:

$$\Delta P_{wind} = \frac{\rho_{oa} \, (S_w \cdot V_H)^2}{2} \tag{15.10}$$

V_H is the local wind speed at the highest point of the building envelope's air barrier and can be calculated from the weather file's V_{wind} using an expression like that of Equation 10.3. S_w is known as the *shelter coefficient*, which has a value between 0.3 (heavy shielding caused by many neighbouring buildings) and 1 (no surrounding objects). This value is usually provided by the user.

The results of Equations 15.7 and 15.8 are then blended to consider the combined impact of the stack effect and wind. Different blending approaches are in use. Additionally, some BPS tools also consider interactions with mechanical ventilation and devices that draw combustion air from the zone.

During this chapter's simulation exercises you will explore which (if any) single-zone models your chosen BPS tool supports. You will also discover how to configure these models and the data you are required to provide.

15.4   NETWORK AIRFLOW MODELS

Network airflow models provide a higher level of modelling resolution than single-zone methods, but demand greater input data from the user. They offer the possibility of calculating the mass flow rate through individual openings in the building envelope. This section builds upon the conceptual basis of network airflow models that was provided in Section 7.3 by describing how they work.

Approach and assumptions

Network airflow models are considered *macroscopic* in that they predict the bulk transport of mass within the building, and between the building and the exterior environment. They can predict air infiltration, natural ventilation, and transfer airflows, as well as airflow interactions with mechanical ventilations systems.

Each zone of the building is treated as a well-mixed volume having uniform conditions and is represented by a single airflow node[ii]. Unlike CFD models they do not predict spatial details, such as airflow patterns within a zone, or the temperature and contaminant distribution within a zone.

The method presumes that mass flows are driven exclusively by pressure differences across openings in the building envelope or through openings connecting zones. Mass balances are formed

---

[ii]A zone could be subdivided and represented by multiple airflow nodes, but this can add substantial complications for the user.

and solved to respect the principle of the conservation of mass. Although the method considers the conservation of mass within the network, the conservation of momentum is not considered. This assumption implies that momentum effects are negligible, which is not true for some airflow situations.

Network airflow models are built upon another important assumption, that hydrostatic conditions prevail. This means that air is considered to be at rest within each zone and that vertical pressure gradients within a zone are caused exclusively by the weight of the column of air between the points.

You will discover the implications of some of these assumptions during this chapter's required reading.

Flow components

Consider the zone represented in Figure 15.3. There are three openings in the envelope: *I* on the left wall, *II* on the right wall, and *III* in the ceiling. The figure illustrates the direction of airflow through these openings at a snapshot in time when air is infiltrating through openings *I* and *II* and exfiltrating through opening *III*. The flow directions may be different at other points in time.

The pressure difference across each opening is determined by local conditions. For example, the exterior pressure at opening *I* (P_e^I) is caused by the combination of the stack effect and wind at that location. This may be very different than the conditions at openings *II* and *III*.

It is up to the user to locate these openings and to characterize them. Indeed, it is up to the user to decide whether there are three openings, 10 openings, or more. The user characterizes each opening using a flow *component* relationship that was first introduced with Equation 7.1. These are simple mathematical expressions that relate the flow rate through the opening to the pressure difference across the opening.

Most BPS tools that include network airflow models provide many such component equations, all of which are empirical in nature. The user might choose, for example, to represent opening *I* using the orifice flow assumption that was discussed in the last section. With this, the mass flow rate through this opening can be

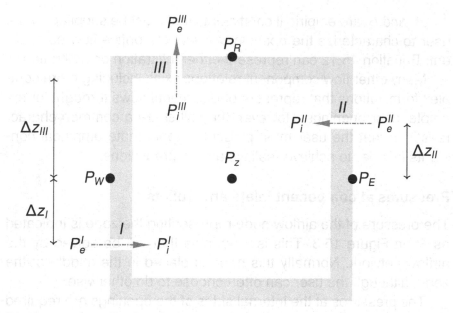

Figure 15.3: Network airflow model representation of a zone with three airflow openings

expressed by:

$$\dot{m}^I_{a,inf} = C_D \cdot A_I \cdot (2\rho)^{1/2} \cdot \text{sgn}\left(P^I_e - P^I_i\right) \cdot \left|P^I_e - P^I_i\right|^{1/2} \quad (15.11)$$

Compared with Equation 15.3 which represented a frictionless orifice, this equation includes the addition of the parameter C_D. Known as a *discharge coefficient*, C_D accounts for friction effects and contraction/expansion effects through the opening. Both it and A_I must be supplied by the user.

It can be seen from Equation 15.11 that $\dot{m}^I_{a,inf}$ could be positive (infiltration) or negative (exfiltration), depending upon the relative magnitudes of the pressures, since the signum function, sgn($P^I_e - P^I_i$) is equal to 1 when $P^I_e > P^I_i$, and equal to -1 when $P^I_e < P^I_i$. Therefore, this component presumes that the resistance to flow is the same in either direction.

Alternatively, the user could choose to represent opening *I* with a component based upon a power law resistance equation:

$$\dot{m}^I_{a,inf} = \mathcal{A}_I \cdot \text{sgn}\left(P^I_e - P^I_i\right) \cdot \left|P^I_e - P^I_i\right|^{\mathcal{B}_I} \quad (15.12)$$

$\mathcal{A}_l$ and $\mathcal{B}_l$ are empirical coefficients that must be supplied by the user to characterize the opening. As with the orifice flow component, Equation 15.12 can represent either infiltration or exfiltration.

Many other flow component relations exist, including more complex formulations that represent bi-directional flows through, for example, door openings. However, they all share a common characteristic in that the user must prescribe appropriate empirical constants in order to achieve realistic airflow predictions.

Pressures at component inlets and outlets

The pressure of the airflow node representing the zone is indicated as P_z in Figure 15.3. This is a variable that will be solved by the airflow network. Normally this node is placed in the middle of the zone, although the user can often choose to do otherwise.

The pressures at the internal sides of the openings are required by the flow component representations, such as Equations 15.11 and 15.12. These pressures can be related to the zone node pressure given the underlying hydrostatic assumption that was mentioned earlier:

$$P_i^l = P_z + \rho_z \cdot g \cdot \Delta z_l$$
$$P_i^{ll} = P_z - \rho_z \cdot g \cdot \Delta z_{ll} \qquad (15.13)$$
$$P_i^{lll} = P_z - \rho_z \cdot g \cdot \Delta z_{lll}$$

where Δz_i are the vertical distances (m) between the zone node and the openings, as indicated in Figure 15.3.

Airflow nodes are also located on the zone's external surfaces. Three such nodes are shown in Figure 15.3. P_E is the pressure at the mid-height[iii] of the east wall (right side of figure), P_W is the pressure at the mid-height of the west wall, and P_R is the pressure on the external surface of the roof. Again, these are variables that will be solved by the airflow network.

The pressures at the external sides of the openings can be related to these nodal pressures using the hydrostatic assumption,

---

[iii] The placement of these nodes is usually at the discretion of the user.

but in this case the density of the outdoor air is used:

$$P_e^I = P_W + \rho_{oa} \cdot g \cdot \Delta z_I$$
$$P_e^{II} = P_E - \rho_{oa} \cdot g \cdot \Delta z_{II} \qquad (15.14)$$
$$P_e^{III} = P_R$$

Wind-driven pressure

Wind may impinge upon some external surfaces, while flowing around others. This causes a unique static pressure at each location of the building envelope. The distribution of these pressures will depend upon wind speed, direction, turbulence, air density, surface orientation, and the surrounding terrain. Generally, pressures will be positive on the windward side of the building, and negative on the leeward side.

These factors are considered in establishing the pressures at the external surface nodes. Normally this is accomplished using an approach based on Bernoulli's flow equation (see Equation 15.2). For example:

$$P_E = P_{atm} + C_P \cdot \rho_{oa} \cdot \frac{V_H^2}{2} \qquad (15.15)$$

P_{atm} is the atmospheric pressure (Pa). V_H is wind speed at height H (usually taken as the height of the airflow node). It is calculated from the weather file's V_{wind} using an expression like that of Equation 10.3. Some tools provide users options for scaling V_{wind} to V_H and for considering the impact of local terrain effects. At this point it is worth recalling the limitations of wind speed data in weather files that were discussed in Section 8.3 and the complications in establishing accurate local wind speed data that were discussed in Section 10.3.

The variable C_P in Equation 15.15 is known as a *pressure coefficient*. It accounts for the wind direction, surface orientation, and terrain effects mentioned above, and has a value in the range $-1 < C_P < 1$. For example, C_P might have a value around 0.7 for windward flow to an exposed wall, approximately -0.3 for leeward flow on a sheltered wall, and as low as -0.8 for flow over a flat roof. Although somewhat analogous to the shelter coefficient (S_w) used in single-zone models (see Equation 15.10), each surface of the building can have a unique value of C_P to consider local flow effects.

Prediction accuracy of network airflow models can depend quite strongly upon C_P because this parameter is sensitive to many factors, such as building shape and height, and the presence of surrounding buildings and objects. It is up to the user to supply a pressure coefficient set appropriate for the building under consideration. Some tools provide facilities for entering data for each surface as a function of wind incidence angle, others use functions to calculate them, while others provide databases of representative pressure coefficient data sets.

Once the pressures at the external nodes are established with the above procedures, the pressures at the external sides of the openings (e.g. P_e^l)—required by the flow component representations, such Equations 15.11 and 15.12—are calculated using Equation 15.14.

Thermal-airflow coupling

Extending the procedures described by Equations 15.11 through 15.15 leads to a set of equations that relate the mass flow rates through openings to the nodal pressures. There will be one equation for each airflow opening defined by the user.

As described in Section 15.1, the zone mass and energy balances depend upon the $\dot{m}_{a,inf}$ mass flows that will result from the solution of these equations. But it can be seen from Equation 15.13 that the airflow network relations also depend upon the energy balances, because ρ_z is a function of T_z.

Dealing with this coupling can complicate the solution procedure outlined in Section 2.8, but numerous approaches have been developed. It is worth noting that some simulation tools implement the network airflow modelling procedure outlined here but without concurrently solving the thermal equations. You will learn about the implications of ignoring the coupling between the airflow and thermal equations through this chapter's required reading.

Modelling air infiltration, natural ventilation, and transfer air

The preceding treatment focused on how network airflow models can predict air infiltration to a zone. This can be extended to include natural ventilation and inter-zone airflows using the exact same procedures.

Consider the floor plan illustrated in Figure 15.4. A user might use flow components to represent natural ventilation through open windows in offices A and B and the meeting room, and other components to represent infiltration through unintended openings in the building envelope in offices C, E, F, and G. For each, the user would be required to provide the empirical constants required by the flow components, such as those appearing in Equations 15.11 and 15.12.

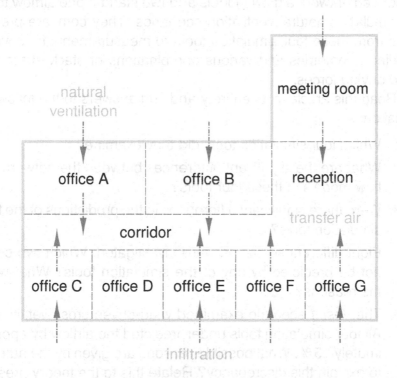

Figure 15.4: Modelling air infiltration, natural ventilation, and transfer air

Components representing open doors could then be specified

for flow between these zones and other zones. The arrows in the figure illustrate the possible flow paths for this scenario (any of these flows could be in the positive or negative direction).

It is important to recognize that this flow network could contain as many or as few components and inter-nodal connections as the user desires. In this example, it can be seen from Figure 15.4 that no components have been defined for office D. As such, during the simulation all $\dot{m}_{a,inf}$ and $\dot{m}_{a,t-in}$ values for this zone will be zero.

15.5 REQUIRED READING

Reading 15–A

Johnson *et al.* (2012) examine the abilities of two BPS tools with integrated network airflow models and two stand-alone airflow tools for simulating natural ventilation scenarios. They compare predictions from these four simulation tools to measurements for several building geometries and various combinations of stack effect and wind driving forces.

Read this article in its entirety and find answers to the following questions:

1. Which four simulation tools did they examine?
2. What are the significant differences between the network airflow models of these four tools?
3. Were there significant differences in the predictions of the four simulation tools?
4. Eight different scenarios were investigated. Which two could not be predicted by any of the simulation tools? What were the reasons?
5. The *Test 1* scenario examined wind-driven cross ventilation. All four simulation tools underpredicted the airflow by approximately 25 %. What possible reasons are given by the authors to explain this discrepancy? Relate this to the theory presented in this chapter's Section 15.4.
6. The *Test 3* scenario examined cross ventilation driven by stack effects. Six combinations of opening sizes and temperature difference between the zone air and the outdoor air

were tested. Which of the six cases was not accurately predicted by any of the simulation tools? Explain why.

15.6 SOURCES FOR FURTHER LEARNING

- Axley (2007) reviews the historical development of multi-zone
  airflow modelling, and describes an alternative to the conventional nodal approach that was described in this chapter.
- Emmerich (2006) applies network airflow modelling to predict
  the performance of natural and hybrid ventilation systems for
  office buildings.
- Wang et al. (2009) describe the empirical validation of two
  single-zone models for predicting the rate of air infiltration to
  16 houses under a wide range of wind and temperature conditions.

15.7 SIMULATION EXERCISES

Revert your BPS tool inputs to represent once again the *Base Case*
described in Section 1.9, including any refinements or corrections
you made following the previous exercises. Perform an annual simulation to produce a fresh set of *Base Case* results for use in the
following exercises.

Exercise 15–A

A constant rate of air infiltration of 0.1 ac/h was prescribed for the
Base Case. This is the second approach listed in Section 15.2.
Now configure you chosen BPS tool to ignore air infiltration (the first
approach) and perform another annual simulation using the same
timestep.

What impact does this change have upon the annual space
heating load? And upon the annual space cooling load? Contrast
this impact to what you observed during the simulation exercises
from previous chapters.

Comment on the implications of ignoring air infiltration or prescribing time-invariant values. Think of some situations where these
approaches would be inappropriate.

Exercise 15–B

In this exercise you will explore calculating air infiltration rates using a single-zone model (the fifth approach listed in Section 15.2).

Now consider that the building's air leakage was characterized with depressurization testing and it was found that $C = 0.01 \, m^3/s \, Pa^n$ and $n = 0.67$.

Calculate the building's effective leakage area using the procedures described in Section 15.3. Then calculate the expected air infiltration rate in ac/h when $\Delta P = 1 \, Pa$. And when $\Delta P = 10 \, Pa$.

Making any necessary assumptions, now calculate the pressure difference caused by the stack effect acting in isolation when $T_{oa} = -20\,°C$. Then calculate the pressure difference caused by wind acting in isolation when $V_{wind} = 5 \, m/s$, again making necessary and appropriate assumptions.

Describe how the *Base Case*'s air infiltration rate would vary if such a model were employed. What impact do you think this would have upon simulation predictions?

Exercise 15–C

Does your chosen BPS tool support a single-zone model for calculating air infiltration? Which models are available? Describe them. Consult your tool's help file or technical documentation to determine how to use this facility. What input data are required?

Now configure your BPS tool to calculate air infiltration using a single-zone model. Use the C and n coefficients given in Exercise 15–B. Perform another annual simulation using the same timestep.

Create a graph that plots the air infiltration rate calculated by your BPS tool versus time for March 6. How does the air infiltration rate vary over this day? When is it highest? Lowest? Explain this trend by examining the weather file's wind speed and outdoor air temperature data for this day.

Exercise 15–D

What is the annual space heating load of the simulation conducted in Exercise 15–C? And the annual space cooling load? How do these results compare with those of the *Base Case*?

What assumptions were you required to make to configure your BPS tool for this simulation? What are the ranges of plausible inputs for each of the parameters that you assumed?

Perform some additional simulations in which you vary some of these parameters over their plausible ranges. How sensitive are the predictions of annual space heating load and annual space cooling load to these parameters? What sources of information or data could you consult to establish appropriate parameters?

Exercise 15–E

Does your chosen BPS tool have a network airflow model? If so, consult your tool's help file or technical documentation to determine how to use this facility. What input data are required from the user?

Now configure your BPS tool to calculate air infiltration using its network airflow model. Make use of the building's air leakage characteristics (the *C* and *n* coefficients) given in Exercise 15–B. Based on this information you will have to make assumptions about the locations and characteristics of the openings in the building envelope.

You may choose to locate a single opening on each wall, or two openings (one high and one low) on each wall and some openings in the floor and roof, or some other configuration. (The possibilities are almost limitless.) Choose appropriate flow components from the options offered by your tool, and establish their input values based on the information you have at hand.

Perform an annual simulation using the same timestep and then examine the airflow results for March 6. How do the directions and magnitudes of the airflows through the components vary over this day? When are airflows the highest? And the lowest? Is air always infiltrating through some components, and always exfiltrating through others, or does this vary over the day? Explain these results by examining the weather file's wind speed, wind direction, and outdoor air temperature data for this day.

Exercise 15–F

What is the annual space heating load of the simulation conducted in Exercise 15–E? And the annual space cooling load? How do these results compare with those of the *Base Case* and those of Exercise 15–C and Exercise 15–D?

What assumptions were you required to make to configure your BPS tool for this simulation? What are the ranges of plausible inputs for each of the parameters that you assumed?

Perform some additional simulations in which you vary some of these parameters over their plausible ranges. How sensitive are the predictions of annual space heating load and annual space cooling load to these parameters? What sources of information or data could you consult to establish appropriate parameters?

15.8   CLOSING REMARKS

This chapter described the possibilities for treating air infiltration and natural ventilation. Through the simulation exercises you saw the significance of air infiltration on simulation predictions and realized the implications of ignoring this mass flow rate, or simply treating it using prescribed values.

The two modelling approaches that can be used for calculating air infiltration as a function of prevailing weather conditions were described. The single-zone method is only applicable for low-rise buildings and relies on user-provided empirical data to describe the building's leakage characteristics. Greater modelling flexibility and resolution are possible with network airflow models, and these can be used to calculate natural ventilation and inter-zone airflow (transfer air) in addition to air infiltration. However, they demand considerable input from the user to define and characterize all possible flow paths. And it was explained that airflow is constrained to these predefined paths.

This completes our treatment of heat and mass transfer processes occurring through the building envelope. Our attention now turns to HVAC systems.

V

HVAC

CHAPTER 16

HVAC distribution systems

THIS part of the book treats mechanical systems that provide heating, ventilation, and air-conditioning (HVAC). The focus of the current chapter is on modelling the components that distribute these services to the building zones.

Chapter learning objectives

1. Realize the differences between idealized, component, and system modelling approaches.
2. Learn how to distinguish between first principle, quasi-first principle, and empirical models.
3. Understand the benefits and complexities in considering transient effects in HVAC component models.
4. Appreciate the complexities and limitations in representing controls.
5. Learn how to configure the various approaches available in your chosen BPS tool.

16.1   HVAC SYSTEM CONFIGURATION AND CONTROL

The possibilities for the configuration of mechanical systems to heat, ventilate, and air-condition buildings are myriad. Some HVAC systems use air as the medium for distributing heating and cooling to occupied spaces. These modulate the temperature, humidity, and/or flow rate of air such that comfort conditions are achieved when the air supplied by the HVAC system is mixed with the room air. The air can be conditioned (heated, cooled, dehumidified) within the ductwork using hydronic coils that are supplied with streams of heated or chilled water. Alternatively, the air can be conditioned by direct heating systems such as furnaces or by direct-expansion refrigeration devices.

Different arrangements of fans and ducting are used with these air-based systems. Some vary the airflow rate in response to thermal conditions (variable air volume, VAV) while others maintain a constant airflow rate but vary the supply temperature (constant volume). There are dual-duct arrangements wherein separate streams of heated and cooled air are supplied and mixed just before delivery to the occupied spaces, and also single-duct arrangements.

Other HVAC systems supply air for ventilation purposes, but accomplish heating and cooling using radiators, radiant floors, chilled beams, heat pumps, fan-coils, or other hydronic devices located within the occupied spaces (often called *terminal devices*). These terminal devices are typically supplied with heated or chilled water streams, but other direct arrangements are possible as described above.

There are many hybrid arrangements of air-based and hydronic-based systems. One common example is a VAV system that provides ventilation, cooling, dehumidification, and partial heating, with the balance of the heating provided by hydronic terminal devices. One possible configuration of such a system is illustrated schematically in Figure 16.1. Some components are omitted for the sake of clarity (e.g. filters). Many variants of such a system exist, including a different ordering of the heating coil, cooling coil, and

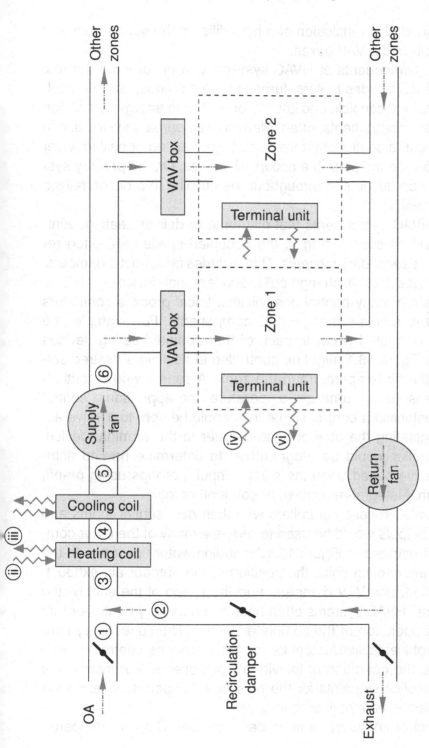

Figure 16.1: Schematic representation of a variable air volume system with terminal heating (air state points identified by Arabic numerals, airflows by ----▶, water state points by Roman numerals[a], and water flows by ∿∿▶)

[a]Numbering is not sequential in order to avoid confusion with the subscripts i and v that are used in other chapters.

supply fan, and the inclusion of a humidifier in the supply duct and reheat coils in the VAV boxes.

Some components of HVAC systems convert or store thermal energy. This includes boilers, furnaces, heat pumps, chillers, cooling towers, and sensible and latent stores. These energy conversion and storage components, often referred to as *primary systems*, can be located in a central plant that distributes heated or chilled water and/or conditioned air to the occupied spaces. Or, the primary systems may be distributed throughout the building in a decentralized arrangement.

The HVAC components that distribute or deliver heating, ventilation, and air-conditioning to the occupied spaces are often referred to as *secondary systems*. This includes fans, ducts, dampers, pumps, pipes, heat exchange coils, and terminal devices.

There are many control possibilities. Local process controllers are used to regulate many HVAC components. For example, the flow rate of heated water to each of the terminal heating devices shown in Figure 16.1 might be controlled to achieve a desired setpoint of the air temperature in the zone. A sensor would continuously measure the zone air temperature and apply some control logic to establish a control signal that would be sent to a valve actuator regulating the flow of heated water to the terminal device. The controller would be programmed to determine how to actuate the valve based upon the sensor input, perhaps using on-off, proportional-integral-derivative, or some other logic.

Separate process controllers with their own sensor inputs and actuator outputs would be used to regulate many of the other components illustrated in Figure 16.1, including water flow rates to the heating and cooling coils, the position of the outdoor air, exhaust, recirculation, and VAV dampers, and the speed of the supply and return fans. HVAC systems often have a supervisory controller that oversees operation of the complete system. The supervisory controller might establish setpoints for local process controllers and sequence the operation of individual components. For instance, it might establish setpoints for the fans and dampers to operate the VAV system in economizer mode.

Control of HVAC systems is never perfect. Setpoint temperat-

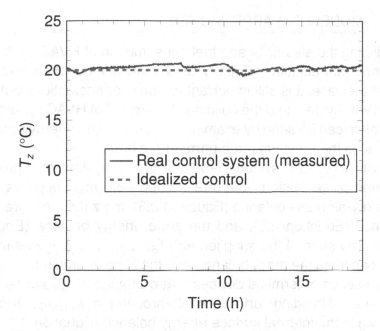

Figure 16.2: Comparison of real and idealized control

ures may not be achieved due to capacity limitations, and overshooting and undershooting can occur due to system, sensor, and actuator transients, as well as due to non-optimized control logic. An example is provided in Figure 16.2. This plots the measured air temperature of a zone heated with a radiant floor that is supplied with hot water through a slow reacting valve that is actuated by an on-off controller. There are numerous overshoots caused by on-off control deadbands, system disturbances such as solar and internal gains, and transients in the radiant floor and valve actuator.

Many factors influence the choice of HVAC system configuration and control, including building type, size, and vintage. Climate and regional traditions are also important determinants. BPS users should be familiar with the possible system configurations and control scenarios because the treatment of HVAC systems is an important aspect in most BPS analyses.

16.2 MODELLING APPROACHES

Predicting the electricity and fuel consumption of HVAC systems is an important objective of many BPS analyses. But even when this is not the case, it is still important to consider interactions between the HVAC system and the building. The import of HVAC system performance can be seen by examining the mass and energy balances studied in the previous three parts of the book.

With air-based systems—or any HVAC system that provides mechanical ventilation—the rate of supply air ($\dot{m}_{SA}$) appears in the zone dry-air mass balance (Equation 2.3), the zone moisture mass balance (Equation 2.6), and the zone energy balance (Equation 2.13). The state of the supplied air (T_{SA}, ω_{SA}, $c'_{P_{SA}}$) figures in both the zone moisture mass balance and the zone energy balance. And the presence of terminal devices that exchange longwave radiation with internal building surfaces is felt through the $q_{lw,HVAC \to i}$ term appearing in the internal surface energy balance (Equation 2.14).

A wide range of methods have been developed to represent HVAC systems and their interaction with the building. Some BPS tools support multiple approaches, which allows users to conduct simulations at various levels of resolution. The types of methods in common use are introduced in this section.

Many BPS tools allow HVAC systems to be represented in an *idealized* manner. With this approach, there is no attempt to represent the performance of individual HVAC components (fans, heat exchange coils, heat pumps, etc.) or to calculate their consumption of electricity and fuel. Rather, the HVAC system's impact on the building is approximated. These methods are described in Section 16.3.

The performance of HVAC systems can be more realistically predicted when the components of HVAC systems are explicitly represented. Some BPS tools use a *component* approach. With this, the user specifies models to represent each significant component and connects these into a network to represent the complete HVAC system. To represent the VAV system depicted in Figure 16.1, the user would have to select models for the fans, heating and cooling coils, dampers, VAV boxes, and terminal devices. Each of these models would require parameters (e.g. empirical constants, nominal electrical power draws, performance at rated conditions). The user would then interconnect the components to represent the possible flow paths determined by the ducting.

Controls would also have to be specified for each component. Some tools attempt to explicitly represent the types of process controls described in the previous section, while others treat the controls in a more idealized fashion. In either case, the user would have to provide parameters for each controller to define sensors, actuators, control logic, and setpoints.

BPS tools that take this component approach offer great flexibility, but demand significant data input and work from the user. Other BPS tools take a *systems* approach for treating HVAC. These offer the user the choice of preconfigured templates that represent many of the common HVAC arrangements, such as the system depicted in Figure 16.1. Usually the user can specify parameters, such as heating and cooling capacities, nominal airflow rates, and nominal fan power draws. The systems approach is less flexible because it is limited to a set of preconfigured templates and control strategies. However, it significantly reduces the amount of work (and the potential for input errors) associated with defining HVAC systems.

With both component and systems approaches, individual models are usually used to represent each component of the HVAC system. Wright (2019) provides a useful scheme for classifying these component models. Some are formulated using mass balances and energy balances based on the first law of thermodynamics. These so-called *first principle models* always employ some simplifying assumptions but they do not rely upon empirical parameters.

In contrast, *empirical models* make no attempt to characterize the internal processes occurring within components. Instead, they use methods such as curve fits to manufacturer or test data to characterize the performance of a complete component. There are also *quasi-first principle models*. The functional form of these is derived from first principles, but they require some empirical parameters.

Models can be further classified as *transient* (dynamic) or *steady-state*. In reality all processes are transient, although some can be approximated as steady-state when they occur at time scales that are much shorter than the timesteps used in a BPS simulation. Transient models are sometimes used to represent HVAC components. With these, the model depends on states from previous timesteps of a simulation. In contrast, many HVAC models ignore these transients. With these, model outputs do not depend upon states from previous timesteps of a simulation. These are sometimes referred to as *quasi-steady-state* models or *algebraic* models.

The following sections illustrate these various modelling approaches. We start with an examination of idealized methods.

16.3   IDEALIZED METHODS

Idealized methods are supported by many BPS tools and are usually the simplest and least time-consuming for the user to configure. It is quite likely that you have employed such an approach for conducting the simulation exercises for the previous chapters. The performance of individual HVAC components (fans, heat exchange coils, heat pumps, etc.) is not considered. Rather, the HVAC system's impact on the building is approximated by calculating $\dot{m}_{SA}$ and T_{SA} (or some surrogate) to represent the heating provision and sensible cooling extraction from a zone using an air-based HVAC system.

With most idealized approaches, the HVAC system is represented as having infinite modulating capability with perfect control. In this way, the exact amount of energy is added (or extracted) to satisfy the setpoint temperature prescribed by the user. With this, there is no overshooting or undershooting of setpoint temperatures. This is illustrated in Figure 16.2 where it can be seen that the idealized HVAC system is able to perfectly satisfy a zone air setpoint temperature of 20 °C by injecting the exact amount of required heating at each timestep of the simulation.

Some tools support the option of providing ventilation air (with or without heat recovery). When configured without ventilation (with some tools this is the only available approach) the idealized system represents a convective heating device located within the zone.

With most tools the user can place capacity limits to investigate the impact of, for example, an undersized cooling system.

Some tools can also represent ideal humidification and dehumidification. With this, the tool also calculates the values of ω_{SA} and $c'_{P_{SA}}$ required to exactly meet the humidity setpoint prescribed by the user.

There is also the possibility to ideally represent terminal devices with some BPS tools. With this, the user can prescribe the fraction of heating and sensible cooling that is provided radiatively and the fraction that is provided convectively. For example, if the user specifies that a terminal device delivers all its heat radiatively (not a realistic scenario), the required heating would be represented by the $q_{lw,HVAC \to i}$ term and $\dot{m}_{SA}$ would be set to zero. In this way, all of the heat added by the HVAC system would appear in the energy balances of the internal surfaces and none would appear in the zone energy balance. Some BPS tools also provide models that allow heat injection or extraction to occur within opaque envelope assemblies. This could be used to represent radiant floor heating or chilled beams in an idealized fashion.

These idealized approaches are usually quite simple to configure and can be extremely useful for many analyses. But it is important to realize that they represent an idealization of how the HVAC system will perform in reality. Moreover, they do not predict the electricity and fuel consumed by the HVAC system, but rather approximate the amount of energy that must be provided or extracted by the HVAC system to maintain conditions within the zones.

The performance of HVAC systems can be more realistically predicted when the components of HVAC systems are explicitly represented. The following sections illustrate the types of models that are used to explicitly represent HVAC components by focusing on the VAV system of Figure 16.1.

16.4   AIR-MIXING DUCT MODELS

Consider the section of the duct in Figure 16.1 that mixes the outdoor air (indicated by state 1) and recirculation air (state 2) upstream of the heating coil (state 3). This mixing process can be represented with a first principle model based on mass and energy balances if a number of simplifying assumptions are made.

If the mixing process is at steady-state and if air leakage from the ducts is ignored, a mass balance on the dry air component of the moist air mixing process can be written as:

$$\dot{m}_{a,1} + \dot{m}_{a,2} = \dot{m}_{a,3} \qquad (16.1)$$

By drawing on the methods introduced in Section 2.5, a mass balance on the water vapour component of the moist air process can be written as:

$$\dot{m}_{a,1} \cdot \omega_1 + \dot{m}_{a,2} \cdot \omega_2 = \dot{m}_{a,3} \cdot \omega_3 \qquad (16.2)$$

If friction and heat losses are ignored, and if kinetic and potential energy effects are neglected, a first law energy balance on the mixing process can be written as:

$$\dot{m}_{a,1} \cdot (h_a + \omega h_v)_1 + \dot{m}_{a,2} \cdot (h_a + \omega h_v)_2 = \dot{m}_{a,3} \cdot (h_a + \omega h_v)_3$$
$$\dot{m}_{a,1} \cdot h_1' + \dot{m}_{a,2} \cdot h_2' = \dot{m}_{a,3} \cdot h_3' \qquad (16.3)$$

where the ' symbols indicate moist air properties (refer to Section 2.6).

Equations 16.1 through 16.3 fully characterize the mixing process subject to these simplifying assumptions. Such approaches are commonly used in BPS tools to treat mixing and diverting components for moist air and liquid streams. However, very few other HVAC components can be treated with first principle methods.

16.5   HEATING AND COOLING COIL MODELS

Many BPS tools represent heating and cooling coils using quasi-first principle models that are based upon the Number of Transfer Units (NTU) heat exchanger method. This is illustrated by focusing on the heating coil of Figure 16.1 which heats the incoming air stream from state 3 to state 4. Heating is accomplished by supplying the coil with a stream of hot water (state ii) that transfers energy to the air as the water is cooled to state iii.

The model described by Wetter (1999) is illustrative of this type of approach. It treats the heat exchanger as being at steady-state, ignores stray heat loss to the surrounding environment, and neglects the thermal resistance of the heat exchanger's solid materials. By neglecting kinetic and potential energy effects and the pressure drop across the coil, and by treating the specific heats as constant, a first law energy balance can be written for the heat transfer from water to air:

$$q_{HX} = \left(\dot{m}c_P'\right)_a \cdot (T_4 - T_3)$$
$$= (\dot{m}c_P)_w \cdot (T_{ii} - T_{iii}) \tag{16.4}$$

where q_{HX} is the rate of heat transfer (W). Conditions on the air side of the heat exchanger are indicated by $_a$ while those on the water side by $_w$. c_P' is the specific heat of moist air (refer to Section 2.6).

Section 16.1 described some of the possibilities for controlling the VAV system depicted in Figure 16.1. It explained that both the flow rate of air and the flow rate of water through the heating coil could be modulated. At any given timestep of the simulation, the model must determine which flow stream has the lower heat capacity rate, and which the higher given this control:

$$(\dot{m}c_P)_{min} = \min \left[\left(\dot{m}c_P'\right)_a, (\dot{m}c_P)_w\right] \tag{16.5}$$

$$(\dot{m}c_P)_{max} = \max \left[\left(\dot{m}c_P'\right)_a, (\dot{m}c_P)_w\right] \tag{16.6}$$

The ratio of Equations 16.5 and 16.6 is also calculated each timestep of the simulation:

$$Z = \frac{(\dot{m}c_P)_{min}}{(\dot{m}c_P)_{max}} \tag{16.7}$$

where Z is known as the *heat capacity ratio*.

The heat exchanger effectiveness is defined to be the actual heat transfer rate relative to the maximum possible heat transfer rate for the prevailing conditions. The maximum possible rate of heat transfer is determined by the stream with the lowest heat capacity rate. It occurs when either the air is heated to the incoming water temperature, or when the water is cooled to the incoming air temperature:

$$\epsilon_{HX} = \frac{q_{HX}}{q_{HX,max}}$$

$$= \frac{q_{HX}}{(\dot{m}c_P)_{min} \cdot (T_{ii} - T_3)} \quad (16.8)$$

ϵ_{HX} depends on the heat exchanger's design, but it is also a function of the flow rates and temperatures of the water and air streams. As such, it must be recalculated each timestep of the simulation.

NTU is defined to be the product of the heat exchange area and the heat exchanger's overall heat transfer coefficient divided by the minimum heat capacity rate:

$$NTU = \frac{(UA)_{HX}}{(\dot{m}c_P)_{min}} \quad (16.9)$$

where U is the heat exchanger's overall heat transfer coefficient (W/m^2 K) and A is the heat exchange area (m^2).

$(UA)_{HX}$ varies with the heat exchanger's operating conditions. The user prescribes a value at the rated water and airflow rates and rated entering water and air temperatures. The user also specifies the ratio of the convective heat transfer coefficients to the air and water streams at these rated conditions. During each timestep of the simulation, the model calculates the $(UA)_{HX}$ value for the prevailing water flow rate and airflow rate, and entering water and air temperatures using these parameters and a series of empirical relations.

Finally, another empirical relation is used to calculate ϵ_{HX} from the results of Equations 16.7 and 16.9. The form of this empirical relation depends upon the heat exchanger configuration. For example, the following form is used for cross flow designs (such as that shown in Figure 16.1):

$$\epsilon_{HX} = 1 - \exp\left[\frac{\exp\left(-Z \cdot NTU^{0.78}\right) - 1}{Z \cdot NTU^{-0.22}}\right] \quad (16.10)$$

Once ϵ_{HX} has been established with Equation 16.10, the rate of heat exchange can be determined with Equation 16.8. And the temperatures of the water and air streams exiting the heat exchanger can be determined with Equation 16.4.

Numerous variations on this approach are used by BPS tools. Some use different empirical relations in place of Equation 16.10 and different empirical approaches for determining $(UA)_{HX}$. Some have added complexity to deal with condensation of water vapour from the air stream for cooling coils. Others drop the assumption of steady-state by considering the thermal resistance and the mass of the heat exchanger's solid materials.

16.6 VARIABLE-SPEED FAN MODELS

The speed of the supply air fan depicted in Figure 16.1 will be modulated in response to the zone's heating or cooling loads. Some control model will be used to establish an appropriate mass flow rate ($\dot{m}_a$) for each timestep of the simulation. It is common for BPS tools to calculate the rate of electricity consumption by the fan's motor for the given $\dot{m}_a$ using an empirical approach, and the state at the fan's exit (state 6) using a quasi-first principle method.

One common approach uses a fourth-order polynomial empirical relationship to determine the rate of electricity consumption:

$$\dot{P}_{el} = \dot{P}_{el}^{rated} \cdot \left[C_0 + C_1 \left(\frac{\dot{m}_a}{\dot{m}_a^{rated}} \right) + C_2 \left(\frac{\dot{m}_a}{\dot{m}_a^{rated}} \right)^2 \right.$$
$$\left. + C_3 \left(\frac{\dot{m}_a}{\dot{m}_a^{rated}} \right)^3 + C_4 \left(\frac{\dot{m}_a}{\dot{m}_a^{rated}} \right)^4 \right]$$

(16.11)

where $\dot{P}_{el}$ is the rate of electricity consumption by the motor driving the fan (W) and $\dot{m}_a$ is the mass flow rate of air through the fan (kg/s).

The other variables in Equation 16.11 are parameters that are supplied by the user (some BPS tools provide default values). $\dot{P}_{el}^{rated}$ is the motor's rate of electricity consumption when the fan is operating at its rated (maximum) flow rate. The C_i coefficients are empirical values that express the relationship between electricity consumption and airflow. Some BPS tools may employ polynomials of a different order than Equation 16.11 and some provide default C_i coefficients, but it is common practice to use such an empirical approach.

The state of the air exiting the fan (state point 6 in Figure 16.1) is usually determined by applying a first law energy balance on the air. In addition to treating the fan as operating at steady-state, kinetic and potential energy effects are neglected. With these assumptions, a first law energy balance on the air stream can be written as:

$$W_{fan} + q_{fan+motor \to a} = \dot{m}_a \cdot (h_6' - h_5') \qquad (16.12)$$

W_{fan} represents the rate of mechanical work done by the fan on the air (W) while $q_{fan+motor \to a}$ is the rate of heat transfer to the air (W). This heat transfer can be due to mechanical or electrical irreversibilities in the motor, fan, or the mechanical coupling between the motor and fan. The $'$ symbols in Equation 16.12 indicate moist air properties (refer to Section 2.6).

Many BPS tools use an empirical relationship to calculate the left side of Equation 16.12, often in the form of:

$$W_{fan} + q_{fan+motor \to a} = \dot{P}_{el} \cdot \left[\eta_{motor} + f_{motor \to a} \cdot (1 - \eta_{motor}) \right] \qquad (16.13)$$

η_{motor} is a user-supplied empirical constant (efficiency) indicating the fraction of the electrical energy consumed by the motor that is converted to shaft power. $f_{motor \to a}$ is another user-supplied empirical constant (some BPS tools fix this at 1) indicating the fraction of heat generation by the motor that is added to the air stream.

For a given inlet state (state 5) and user-supplied empirical constants, the model combines Equations 16.11 to 16.13 to fix the state of the air exiting the fan (state 6).

16.7   HYDRONIC RADIATOR MODELS

The flow rate (or temperature) of the heated water supplied to the terminal heaters depicted in Figure 16.1 will be modulated in response to the zone's heating load. Some control model will be used to establish an appropriate mass flow rate ($\dot{m}_w$) and supply temperature (T_{iv}) for each timestep of the simulation. There is considerable disparity in the models used by BPS tools to treat the terminal devices receiving this hot water. We consider here approaches used for one type of terminal device, hydronic radiators, to illustrate the range of methods in use.

Steady-state approaches were used for all of the component models described in the previous sections because those components have relatively low mass, and therefore respond at time scales that can be considered short relative to typical simulation timesteps. However, this assumption is less appropriate for hydronic radiators, which are fabricated with heavy gauge metals and which encapsulate a significant mass of water within their internal piping.

Some models are structured to consider the impact of thermal transients resulting from this amount of mass. One example is that described by Hensen (1991). This model requires the user to specify the combined mass of the radiator's solid materials and its encapsulated water, as well as the device's average specific heat. This mass is subdivided into a number of segments, each of which is treated as being at a uniform temperature. Water entering the radiator flows sequentially from one segment to the next, exiting the radiator from the last segment. This approach is illustrated in Figure 16.3. The radiator's supply (iv) and return (vi) states correspond to those in Figure 16.1.

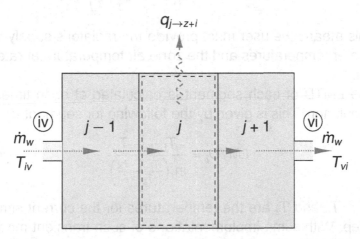

Figure 16.3: Radiator discretized into a number of segments (the control volume [- - - - -] encompasses the solid mass and water contained within the j'th segment)

A first law energy balance is formed for each segment by ignoring friction, kinetic and potential energy effects, and by treating the water as incompressible with constant specific heats. Referring to Figure 16.3, the resulting energy balance for segment j is given by:

$$(mc_P)_j \cdot \frac{dT_j}{dt} = -q_{j \rightarrow z+i} + (\dot{m}c_P)_w \cdot (T_{j-1} - T_j) \tag{16.14}$$

where j represents the segment under consideration and $j-1$ the upstream segment. Segment j is assumed to be at the uniform temperature of T_j.

The left side of Equation 16.14 represents the transient storage of energy by segment j. The rate of heat output by convection and longwave radiation from segment j is given by $q_{j \rightarrow z+i}$. The convection portion (z) will be reflected in the zone energy balance while the radiative portion (i) will be distributed to the zone's internal surfaces.

At each timestep of the simulation each segment's heat output is calculated using a *log mean temperature difference* (LMTD) approach. The LMTD of the radiator at its rated conditions is calculated based upon user inputs:

$$LMTD_{rad}^{rated} = \frac{T_{iv}^{rated} - T_{vi}^{rated}}{\ln\left(\frac{T_{iv}^{rated} - T_z^{rated}}{T_{vi}^{rated} - T_z^{rated}}\right)} \tag{16.15}$$

This means the user must provide the radiator's supply and return water temperatures and the zone air temperature at rated conditions.

The LMTD of each segment is calculated at each timestep of the simulation. This is given by the following for segment j:

$$LMTD_j = \frac{T_{j-1} - T_j}{\ln\left(\frac{T_{j-1} - T_z}{T_j - T_z}\right)} \tag{16.16}$$

T_{j-1}, T_j, and T_z are the temperatures for the current simulation timestep. With some implementations of such transient models, T_z can be replaced by some weighted average of T_z and the temperatures of the zone's internal surfaces. (This requires the user to specify which internal surfaces to consider, effectively indicating the radiator's placement within the zone.)

The heat output is then calculated for each of the segments:

$$q_{j \to z+i} = \frac{q_{rad \to z+i}^{rated}}{J} \cdot \left[\frac{LMTD_j}{LMTD_{rated}} \right]^n \tag{16.17}$$

where J is the number of segments used to represent the radiator. $q_{rad \to z+i}^{rated}$ is the radiator's heat output at the rated conditions, another parameter that must be supplied by the user. The user must also provide the exponent n (an empirical constant) for Equation 16.17.

The total rate of heat transfer from the radiator to the containing zone can then be determined by summing the contributions of each segment:

$$q_{rad \to z+i} = \sum_{j=1}^{J} q_{j \to z+i} \tag{16.18}$$

Some BPS tools treat $q_{rad \to z+i}$ as a convective heat transfer to the zone, while others treat it as a radiative heat transfer to the zone's internal surfaces. In some cases the user can provide additional information to apportion $q_{rad \to z+i}$ into radiant and convective portions. In this way, some of the radiator's output could be represented using the $q_{lw,HVAC \to i}$ in the internal surface energy balance (Equation 2.14) and some using the $\dot{m}_{SA}$ and T_{SA} terms (or some surrogate) in the the zone energy balance (Equation 2.13).

This can be considered a quasi-first principle method. As can be seen from the above treatment, such transient models demand considerable input from the user. But they offer the advantage of considering the impact of thermal transients on the radiator's heat output. This can be seen in Figure 16.4, which plots the results for a 1-hour period from a simulation conducted with such a model. The radiator's process controller was configured with an on-off control logic to provide a constant flow rate of water at 75 °C to the radiator (state iv) whenever T_z dropped below 19 °C, and to stop the flow of water only when T_z rose above 21 °C. The resulting oscillations in T_z and the on-off control of $\dot{m}_w$ can be seen in the top of the figure. (Contrast the T_z oscillations to the constant temperature of the idealized control data series plotted in Figure 16.2.)

The temperature of the radiator as well as the heat transfer rate from the radiator to the zone can be seen in the bottom of Figure 16.4. When the controller stops the flow of hot water after 25 minutes (top of figure), it can be seen that the temperature of the radiator begins to decay (bottom of figure). Even though the flow of hot water has ceased, the radiator continues to transfer energy to the zone (bottom of figure). Indeed the radiator is continuously heating the zone during this 1-hour period even though the flow of water stops three times. A careful examination of the top of the figure after 25 minutes reveals that T_z continues to increase even after the flow of water has stopped. In a similar fashion, when the controller restarts the flow of water there is a time delay before the radiator's heat output reaches its maximum level (inspect the bottom of the figure after 37 minutes). These types of control overshoots and dynamic responses can only be predicted as a result of the transient nature of the radiator model.

In contrast to the transient model described above, it is quite common for BPS tools to represent heat transfer from radiators using steady-state methods. These have a structure very similar to that of the heating coil model described in Section 16.5. The $(UA)_{HX}$ of the radiator is calculated from user inputs that characterize performance under rated conditions. But unlike with the heating coil model, this value is treated as a constant throughout the simulation.

The radiator's NTU is determined each timestep using Equation 16.9 and the radiator's effectiveness determined with an expression like that of Equation 16.10. The airflow rate over the radiator is assumed in order to perform these calculations. The heat transfer from the radiator to the room can then be determined by:

$$q_{rad \to z+i} = \epsilon_{HX} \cdot (\dot{m}c_P)_{min} \cdot (T_{iv} - T_z) \tag{16.19}$$

As with the transient method described above, additional user-defined inputs can be used to apportion $q_{rad \to z+i}$ into radiant and convective outputs.

Another steady-state approach used by some BPS tools

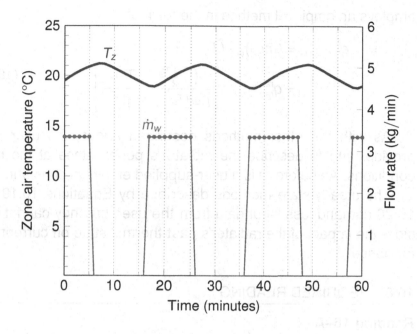

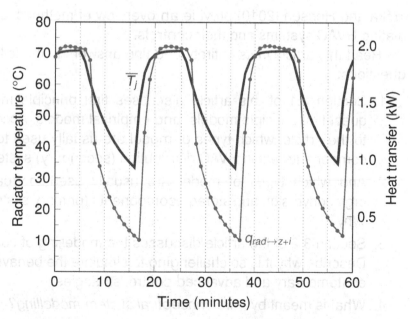

Figure 16.4: Simulation predictions using a quasi-first principle transient radiator model

employs an empirical method in the form of:

$$q_{rad \to z+i} = (\dot{m}c_P)_w \cdot (T_{iv} - T_{vi})$$

$$= q_{rad \to z+i}^{rated} \cdot \left[\frac{\left(\frac{T_{iv}+T_{vi}}{2} - T_z\right)}{\left(\frac{T_{iv}+T_{vi}}{2} - T_z\right)^{rated}} \right]^n \qquad (16.20)$$

As with the other methods described above, the user must provide data to describe the radiator's performance at the rated conditions. As before, n is a user-supplied empirical constant.

The steady-state methods described by Equations 16.19 and 16.20 demand less input data from the user, but they cannot consider the impact of the radiator's past thermal state on current performance.

16.8   REQUIRED READING

Reading 16–A

Trčka and Hensen (2010) provide an overview of methods for simulating HVAC systems and their controls.

Read this article in its entirety and find answers to the following questions:

1. Section 3.1 of this article discusses first principle models, quasi-first principle models, and empirical models. According to the article, which types of model are usually used to treat components within HVAC distribution (secondary) systems?

2. And which types of model are usually used to treat energy conversion and storage components (primary systems)? Why?

3. Section 3.2 of the article discusses the modelling of controls. Describe why it is so challenging to simulate the behaviour of contemporary and advanced control strategies.

4. What is meant by *pure conceptual system modelling*?

5. What is meant by *equation-based system modelling*? List some tools that employ this approach.

6. Based on this article, explain why it is quite challenging to explicitly represent mass and energy flows in HVAC systems.

16.9 SOURCES FOR FURTHER LEARNING

- ASHRAE (2016) thoroughly describes HVAC system configurations and components.
- Wright (2019) describes the structure of HVAC systems models, their calibration, and solution methods.
- Brideau *et al.* (2016b) describe BPS models that have been developed for radiant floor heating systems and present an inter-model comparison of their simulation predictions.
- Wetter (2009) argues the advantages of equation-based modelling approaches for treating HVAC components and control systems.

16.10 SIMULATION EXERCISES

Revert your BPS tool inputs to represent once again the *Base Case* described in Section 1.9, including any refinements or corrections you made following the previous exercises. Perform an annual simulation to produce a fresh set of *Base Case* results for use in the following exercises.

Exercise 16–A

Consult your BPS tool's help file or technical documentation to determine which of the HVAC modelling approaches described in Section 16.2 it supports. Which of these approaches did you employ for the *Base Case*? Describe how this method works by making reference to the material presented in this chapter.

Extract the *Base Case* results for March 6 and plot the zone air temperature versus time. Do your results resemble either of the series plotted in Figure 16.2? Explain why or why not.

Now plot the rate of energy transfer from the HVAC system versus time on this day. How does the rate of heat addition or extraction vary throughout the day? Could an actual HVAC system be controlled to perform in this manner? If not, what implications does this have upon your simulation predictions?

Exercise 16–B

You were told to configure the HVAC system for the *Base Case* with sufficient heating and cooling capacity to maintain the indoor air between 20 °C and 25 °C at all times. How did you configure your chosen BPS tool to accomplish this?

Now reduce the heat addition capacity of the HVAC system to 1 kW and perform another annual simulation. Plot the zone air temperature versus time for the month of January. What do you observe?

Exercise 16–C

Revert to your *Base Case*.

For the *Base Case* you were told to configure the HVAC system such that heat injection and extraction is 100 % convective and 100 % sensible. In this manner the HVAC system's interactions will appear as the $\left(\dot{m}_a \cdot c_P' \right)_{SA} \cdot (T_{SA} - T_z)$ term in the zone energy balance (Equation 2.13).

Are you able to alter this treatment in your chosen BPS tool using an idealized approach? Can the HVAC system be made to inject/extract energy radiatively rather than convectively, or by a mix of convection and radiation? In this way, some or all of the HVAC system's interventions would appear as the $q_{lw,HVAC \to i}$ term in the internal surface energy balance (Equation 2.14). Consult your BPS tool's help file or technical documentation to determine how the $q_{lw,HVAC \to i}$ term is distributed to the zone's internal surfaces.

Now configure the HVAC system to inject/extract entirely or partially by radiation and perform another annual simulation. If you cannot do this with your chosen BPS tool, then explain why by making reference to the material presented in this chapter.

Examine the zone air temperature and internal surface temperature predictions for March 6. Contrast these to the results from the *Base Case*. What impact does this change have upon the temperature predictions?

What impact does this change have upon the annual space heating load? And upon the annual space cooling load?

Exercise 16–D

Revert to your *Base Case*.

Does your chosen BPS tool provide an idealized model for hydronic floor terminal heating? If so, configure it for such a system, making any necessary assumptions and approximations. Conduct an annual simulation.

Extract results for March 6 and plot the zone air temperature versus time. How does this compare with your results from Exercise 16–A?

What impact does this change have upon the annual space heating load? And upon the annual space cooling load? Are these results consistent with your expectations? If not, explain why.

Exercise 16–E

Revert to your *Base Case*.

Consult your BPS tool's help file or technical documentation to determine whether it can explicitly model hydronic radiator terminal heating. Does it include any of the types of models illustrated in Section 16.7? Describe its model.

Configure your tool to represent terminal heating with a hydronic heating system. Choose appropriate models to explicitly represent a radiator with a nominal heat output of ~2 kW that is supplied with water at 90 °C by a pump whenever there is a demand for heat. What other components must you include?

What input data are you required to provide? Consult technical information from manufacturers of radiator and pump products that are available in your region to establish appropriate input values. Make any necessary assumptions regarding the water flow rate, control of the pump, and parameters for the radiator and pump models. Choose an appropriate timestep and conduct another annual simulation.

Extract results for March 6 and plot the zone air temperature versus time. How do your results compare with those from Exercise 16–A? Do your results resemble either of the series plotted in Figure 16.2? Explain why or why not.

What impact does this change have upon the annual space

heating load? And upon the annual space cooling load? Are these results consistent with your expectations?

16.11   CLOSING REMARKS

This chapter described the types of methods that can be used for treating HVAC distribution systems. It explained that many BPS tools allow HVAC systems to be represented with idealized approaches, which are usually quite simple to configure and can be extremely useful for many analyses. However, it is important to be aware that although these approaches can approximate the amount of energy that must be provided or extracted by the HVAC system to maintain conditions within the zones (i.e. heating and cooling loads), they do not calculate the electricity and fuel consumed by the HVAC system.

The distinction between component and system approaches was made, and it was explained that both make use of models to represent the mass and energy flows within HVAC distribution systems. Descriptions of models were provided to illustrate first principle, quasi-first principle, and empirical approaches. Additionally, both transient and steady-state models of hydronic radiators were described to demonstrate both the benefits of transient approaches as well as the complexities they introduce.

Through the required reading and the simulation exercises you became aware of the approaches supported by your chosen BPS tool, and learned how to configure them.

Our attention now turns to the treatment of the energy conversion and storage components of HVAC systems.

CHAPTER 17

Energy conversion and storage systems

T HIS chapter describes models that are used for representing the energy conversion and storage components (primary equipment) of heating, ventilation, and air-conditioning systems.

Chapter learning objectives

1. Understand that steady-state and empirical models are commonly employed by BPS tools for predicting the performance of energy conversion components.

2. Realize some possible sources of empirical data that can be used to establish input parameters.

3. Learn about the assumptions underlying models used for representing energy storage devices.

4. Learn how to configure the various approaches available in your chosen BPS tool.

17.1 MODELLING APPROACHES

The previous chapter introduced a classification scheme for HVAC component models. We saw how first principle, quasi-first principle, empirical, steady-state, and transient approaches could be used to treat HVAC distribution components.

Although models using all of these approaches have been developed for HVAC primary equipment, we will see in the following sections that most BPS tools employ steady-state and empirical models for treating HVAC energy conversion components, and transient quasi-first principle models for energy storage components.

The following sections describe the types of models that are in common usage. This is accomplished by focusing on a few components that collectively demonstrate the spectrum of methods.

17.2 COMBUSTION BOILER MODELS

The previous chapter described models that can represent the heating coil (Section 16.5) and terminal hydronic radiators (Section 16.7) of the VAV system shown in Figure 16.1. These components are supplied by streams of hot water (state points ii and iv in Figure 16.1).

Although not shown in that figure, a hydronic loop including a boiler, circulating pump, valves, and converging and diverging junctions supplies this hot water. Models are also required to represent each of these components. Methods similar to those presented in Section 16.4 could be used to represent the junctions in the hydronic loop, and approaches like those described in Section 16.6 could be used for the pump. We consider now techniques for representing the boiler.

Figure 17.1 schematically represents a boiler. A stream of water flows through the boiler's heat exchanger at a rate of $\dot{m}_w$. This water enters the boiler at a temperature of T_1 and is heated to T_2.

A control volume is drawn around the entire boiler. This includes all internal components as well as the mass of water contained within the boiler's heat exchanger. A first law energy balance can be formed for this system. By neglecting kinetic and potential energy effects and the pressure drop of the flowing water, and by treating

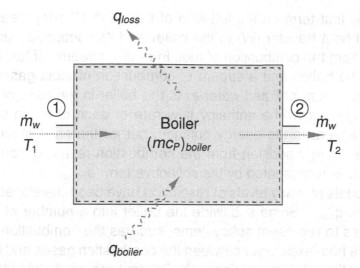

Figure 17.1: Boiler (control volume [-----] encompasses entire boiler and water contained within its heat exchanger)

the specific heats as constant, the energy balance on the boiler's control volume can be written as:

$$q_{boiler} - q_{loss} = (mc_P)_{boiler} \cdot \frac{dT_{boiler}}{dt} + (\dot{m}c_P)_w \cdot (T_2 - T_1) \quad (17.1)$$

The first term on the right side of this equation represents the transient storage of energy by the mass of the boiler and its contained water. Consideration of this term requires the user to provide $(mc_P)_{boiler}$ as an input parameter. The second term on the right side represents the energy added to the water flowing through the boiler's heat exchanger.

The second term on the left side represents the rate of stray heat loss (W) from the boiler to the surrounding room via convection and longwave radiation, and is commonly calculated using an expression like:

$$q_{loss} = (UA)_{boiler} \cdot (T_{boiler} - T_z) \quad (17.2)$$

where $(UA)_{boiler}$ is a heat loss coefficient (W/K, another input parameter required from the user), and T_z is the temperature of the zone containing the boiler.

The first term on the left side of Equation 17.1 represents the rate of heat transfer (W) to the boiler and its contained water resulting from the combustion of fuel. In reality, streams of fuel and air enter the boiler, and a stream of combustion product gases exits. Also, a stream of liquid water exits the boiler in the case of a condensing device. The enthalpy flow rate of each of these streams could appear in the energy balance, but for the sake of simplicity the net energy addition from the combustion resulting from these streams is represented by the collective term q_{boiler}.

Models of many levels of resolution have been developed to determine q_{boiler}. Some subdivide the boiler into a number of control volumes to represent subsystems, such as the combustion chamber, the heat exchanger between the combustion gases and the water, and the combustion air supply. Others further subdivide the heat exchanger to more accurately determine the extent of condensation of water vapour from the combustion gases, and therefore the rate of latent heat transfer. Many such detailed models, which are usually based upon quasi-first principle approaches, can be found in the literature.

However, much simpler empirical models that treat the boiler in its entirety are more commonly employed by BPS tools. These methods usually relate q_{boiler} to the rate of fuel consumption using an efficiency:

$$\dot{m}_{fuel} \cdot HHV = \frac{q_{boiler}}{\eta_{boiler}} \tag{17.3}$$

The mass flow rate of fuel ($\dot{m}_{fuel}$ in kg/s) is a variable whose solution is commonly sought when HVAC systems are simulated. HHV is the fuel's heating value (J/kg), which is either input by the user or defaulted by the BPS tool. The boiler's efficiency is represented by η_{boiler} ($-$). This parameter is sensitive to the boiler's operating conditions, and thus can vary each timestep of the simulation.

Some boilers have multi-stage or fully modulating burners, while others rely on on-off cycling to satisfy thermal loads. The performance of some boilers—especially condensing units—is quite sensitive to the flow rate and temperature of the water stream. Although the functional forms vary from tool to tool, most of these empirical models use some parametric formulation to calculate the efficiency in response to these factors. One example is given by:

$$\eta_{boiler} = C_0 + C_1 \cdot PLR + C_2 \cdot PLR^2$$
$$+ T_2 \cdot \left[C_3 + C_4 \cdot PLR + C_5 \cdot PLR^2 \right] \qquad (17.4)$$

PLR (−), known as the *part load ratio*, relates the boiler's heat output at the current timestep to its rated output (another input parameter supplied by the user):

$$PLR = \frac{q_{boiler}}{q_{boiler}^{rated}} \qquad (17.5)$$

The C_i coefficients of Equation 17.4 are empirical values that must be supplied by the user. These represent the performance of a particular boiler and consider the impact of on-off cycling, burner modulation, condensation of product gases, and other operational factors. With some tools the impact of stray heat losses and transient effects are also encapsulated in these coefficients, in which case the q_{loss} and $\frac{dT_{boiler}}{dt}$ terms are dropped from Equation 17.1.

Some manufacturers publish product specifications that include sufficient data to allow the user to regress Equation 17.4 to determine accurate C_i coefficients. Unfortunately, in many cases the BPS tool user will be unable to find sufficient manufacturer or independent test data to establish accurate C_i coefficients for a particular boiler, necessitating an approximation or reliance upon BPS tool default values representing generic equipment.

Equations 17.1 to 17.5 define the boiler's operation using the commonly employed empirical approach. This set of equations is usually constrained by models representing other components in the HVAC system and their control. For example, other models might fix the values of $\dot{m}_w$ and T_1 for the given timestep, and a control model of the boiler might establish the value of q_{boiler}. Subject to these constraints, the solution of Equations 17.1 to 17.5 would lead to $\dot{m}_{fuel}$ and T_2.

As mentioned, some BPS tools include the transient and heat loss terms, while others represent the impact of these effects in the boiler efficiency. Many alternative functional forms of Equation 17.4 are also in use, although the empirical method outlined here is commonly applied by BPS tools.

17.3 HEAT PUMP MODELS

Energy conversion devices based upon vapour-compression refrigeration cycles (chillers, heat pumps, refrigerators) are most often treated in BPS tools with empirical approaches much like that described in the previous section for boilers.

Consider the water-to-water heat pump and its two connected water streams that are illustrated in Figure 17.2. One water stream transfers energy into the refrigerant at the heat pump's evaporator, while the other transfers energy out of the refrigerant at the condenser.

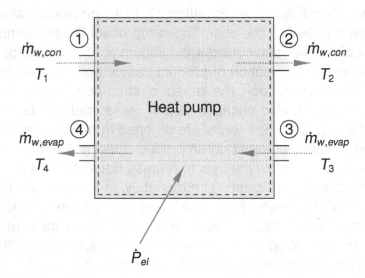

Figure 17.2: Heat pump (control volume [- - - - -] encompasses entire heat pump and water contained within its heat exchangers)

The rate of heating provided by the heat pump can be calculated by focusing on the water stream flowing through the condenser. By making the same assumptions as with the boiler, but in this case neglecting the transient and heat loss terms, the following first law energy balance can be written:

$$q_{HP \to w,con} = (mc_P)_{w,con} \cdot (T_2 - T_1) \tag{17.6}$$

The steady-state empirical models used by many BPS tools

make no attempt to represent the thermodynamic processes occurring within the heat pump. Rather, they employ polynomial equations to determine the heat pump's heating output as a function of its operating conditions. One such model is given by:

$$q_{HP \to w,con} = q_{HP \to w,con}^{rated} \cdot \left[C_0 + C_1 \cdot T_3 + C_2 \cdot T_1 \right.$$
$$\left. + C_3 \cdot \dot{m}_{w,evap} + C_4 \cdot \dot{m}_{w,con} \right] \tag{17.7}$$

Where $q_{HP \to w,con}^{rated}$ is the heat pump's heating capacity at its rated conditions and C_i are empirical constants. All of these values must be supplied by the user. In some cases these can be derived from the published data provided by manufacturers.

With these models the heat pump's electricity consumption is determined in a similar manner:

$$\dot{P}_{el} = \dot{P}_{el}^{rated} \cdot \left[C_0 + C_i \cdot T_3 + C_{ii} \cdot T_1 \right.$$
$$\left. + C_{iii} \cdot \dot{m}_{w,evap} + C_{iv} \cdot \dot{m}_{w,cond} \right] \tag{17.8}$$

where $\dot{P}_{el}^{rated}$ is the electrical power draw at rated conditions and C_i are additional empirical constants, all of which must be supplied by the user as input parameters.

Although they may employ different functional forms than Equations 17.7 and 17.8, many BPS tools employ steady-state empirical models along these lines. Some may include additional elements, such as transient terms and terms to represent stray heat losses from the heat pump.

There are also models in use by BPS tools that explicitly represent the thermodynamic processes within the heat pump cycle. These necessarily apply a number of assumptions about the cycle configuration and demand a considerable number of parameters from the user, such as the chemical composition of the refrigerant, UA values for the evaporator and condenser, the degree of superheat at the compressor inlet, empirical parameters for calculating the compressor's electrical power draw, etc.

17.4 SOLAR THERMAL COLLECTOR MODELS

Many BPS tools employ quasi-first principle models to represent solar thermal collectors. Consider the schematic of a solar collector shown in Figure 17.3. Some of the solar irradiance incident upon the collector will be transmitted through its transparent cover/glazing (many designs exist) and a portion of this energy will be captured by the collector's absorber. Water (usually mixed with an antifreeze) flows through the collector's heat exchanger and is heated from T_1 to T_2 by solar energy that is absorbed by the collector. A portion of the absorbed solar energy will be lost to the exterior environment by convection and longwave radiation, the amount depending upon the collector's design, the flow rate and temperature of the fluid, and environmental conditions.

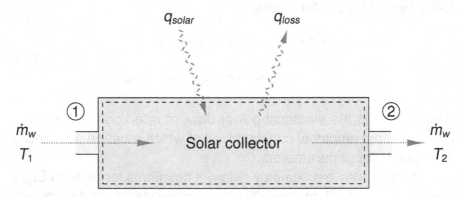

Figure 17.3: Solar thermal collector (control volume [- - - - -] encompasses entire solar thermal collector)

By making the same assumptions as with the boiler and by neglecting the transient term, the following first law energy balance can be written:

$$(mc_P)_w \cdot (T_2 - T_1) = q_{solar} - q_{loss}$$
$$= A_c \cdot \left[(\tau\alpha)_\perp \cdot K_\theta \cdot G_{solar \to e} \right.$$
$$\left. - U_1 \cdot (T_1 - T_{oa}) - U_2 \cdot (T_1 - T_{oa})^2 \right]$$
(17.9)

The left side of Equation 17.9 represents the rate of energy addition to the fluid stream flowing through the solar collector ($_w$ represents the properties of the water and antifreeze mixture). The first term in the square brackets on the right accounts for the rate of absorption of solar irradiance, while the second and third terms represent heat losses to the exterior environment.

A_c is the area of the solar collector (m²) and T_{oa} is the outdoor air temperature (°C or K). $G_{solar \to e}$ is the global incident irradiance on the collector (W/m²), the sum of the solar beam, diffuse, and ground-reflected irradiance. It is determined using the procedures detailed in Chapter 9.

$(\tau\alpha)_\perp$ is the fraction (−) of the incident solar irradiance that is transmitted through the collector's cover and then absorbed (directly or following reflections) by its absorber when the irradiance is perpendicular to the cover. K_θ (−), known as an *incident angle modifier*, modifies $(\tau\alpha)_\perp$ to account for the impact of off-normal angles of incidence. It has a value of unity at normal incidence and higher or lower values at other angles depending upon the geometry and material of the collector's covers and absorber. U_1 (W/m² K) and U_2 (W/m² K²) are heat loss coefficients that account for convection and longwave radiation heat transfer from the collector to the exterior environment.

$(\tau\alpha)_\perp$, U_1, U_2, and K_θ are empirical constants that must be provided by the user. Most solar thermal collectors are subjected to standardized testing that is conducted with solar irradiance simulator lamps under steady-state conditions. It is common for these tests to span a range of temperatures and incidence angles. These test results—which are widely available[i]—can be used to establish the model's empirical parameters.

The values of $\dot{m}_w$, T_1, T_2, T_{oa}, and $G_{solar \to e}$ are measured during a series of tests conducted with a normal angle of incidence for the beam irradiance. These data are regressed to find the quantities of $(\tau\alpha)_\perp$, U_1, and U_2 that best fit Equation 17.9.

K_θ is commonly represented as an empirical function of the

---

[i]Both Solar Keymark and the Solar Rating and Certification Corporation provide extensive databases.

incidence angle:

$$K_\theta = 1 + C_0 \cdot \left[\frac{1}{\cos\theta} - 1 \right] + C_1 \cdot \left[\frac{1}{\cos\theta} - 1 \right]^2 \qquad (17.10)$$

where θ is the incidence angle (refer to Section 9.3).

Data from additional tests performed at off-normal angles of incidence are regressed to find the values of C_i that best fit Equation 17.10.

Many BPS tools provide quasi-first principle models of this type, but there are many variations. Some tools use different functional forms than Equation 17.10 and calculate separate values of incidence angle modifiers for the beam, sky diffuse, and ground-reflected irradiance. Also, models for evacuated tube collectors usually provide separate incidence angle modifiers for the transverse and longitudinal directions.

Many tools apply a factor to Equation 17.9 to account for variations in $\dot{m}_w$, and some subdivide the collector into a number of segments in series to improve the calculation of heat losses. Alternate forms to Equation 17.9 that calculate the heat losses using T_2 rather than T_1, or an average of the two are sometimes used. Some tools also add a time derivative term like that of the boiler model of Equation 17.1 to account for transient effects.

17.5 THERMAL ENERGY STORAGE MODELS

Water tanks are the most prevalent type of thermal store used in HVAC systems. They can be used to buffer between intermittent energy supplies and building thermal demands, to enable longer operational cycles for slow-responding equipment, or to temporally shift energy consumption.

Many tanks are designed to encourage stratification such that the water near the top of the tank can be significantly warmer than that stored near the bottom. This is advantageous for the performance of some heating equipment, such as heat pumps and solar thermal collectors (see Equations 17.8 and 17.9).

Of the many BPS tools that provide explicit models of stratified storage tanks, most assume a one-dimensional temperature distribution in the vertical direction. This approach is illustrated in Figure

17.4, which shows a tank subdivided into six strata (the number of strata is usually specified by the user). Each stratum is assumed to be at a uniform temperature. In the simple example shown here there is a single inlet located near the top of the tank, and a single outlet located near the bottom. Typically the user would specify the number of inlets and outlets and their locations.

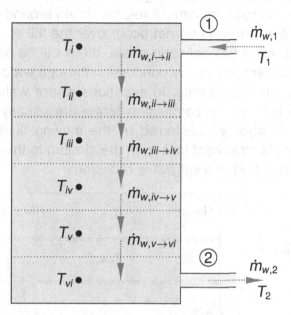

Figure 17.4: Water tank with single inlet port and single outlet port represented with six strata

In reality, the momentum and turbulence of the water stream at the inlet will impact water flow patterns within the tank. Flow entrainment, molecular and turbulent diffusion, and the tank's geometrical details can all have an impact. However, no attempt is made to resolve these details with this modelling approach. Rather, the water is idealized to flow in the manner illustrated in Figure 17.4. Each of the inter-strata flows (e.g. $\dot{m}_{w,i \to ii}$) indicated in the figure represents the net mass exchange between two strata.

Mass and energy balances are formed for each stratum. If the water is treated as incompressible, then the mass of the water contained in each stratum remains constant. With this assumption and for the simple configuration depicted in Figure 17.4, a mass balance leads to:

$$\dot{m}_{w,1} = \dot{m}_{w,i\rightarrow ii} = \dot{m}_{w,ii\rightarrow iii} = \dot{m}_{w,iii\rightarrow iv}$$
$$= \dot{m}_{w,iv\rightarrow v} = \dot{m}_{w,v\rightarrow vi} = \dot{m}_{w,2} \qquad (17.11)$$

The method of forming energy balances is described by focusing on stratum ii, which is illustrated in Figure 17.5. The stratum's net mass exchanges with the neighbouring strata are shown, as are heat transfers across the control volume drawn around the stratum.

Since the water flows do not occur over the full cross-sectional area of the boundaries between strata, there will be heat transfers over the quiescent regions by conduction through water. These are indicated by the q_{cond} terms. In situations where water is flowing through the tank, these conduction terms are usually quite small relative to the energy transferred by the moving fluid. The $q_{loss,ii}$ term represents stray heat loss from the stratum to the surrounding room via convection and longwave radiation.

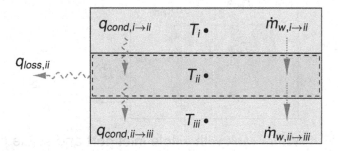

Figure 17.5: Mass and energy flows for a single stratum of the tank

By making the same assumptions as with the boiler, a first law energy balance on the stratum can be written as:

$$[q_{cond,i\rightarrow ii} - q_{cond,ii\rightarrow iii}]$$
$$+ [\dot{m}_{w,i\rightarrow ii} \cdot c_{P,i} \cdot (T_i - T_{ii}) - \dot{m}_{w,ii\rightarrow iii} \cdot c_{P,ii} \cdot (T_{ii} - T_{iii})] \qquad (17.12)$$
$$= (mc_P)_{ii} \cdot \frac{dT_{ii}}{dt} + q_{loss,ii}$$

The conduction terms are commonly approximated using the same techniques that were described in Section 13.3 (see Equation 13.8):

$$q_{cond,i\rightarrow ii} \approx k_w \cdot A_{tank} \cdot \frac{T_i - T_{ii}}{\Delta_{strata}} \qquad (17.13)$$

where k_w is the conductivity of water (W/mK), A_{tank} is the cross-sectional area of the tank (m^2), and Δ_{strata} is the thickness of the stratum (m).

The techniques that were also described in Section 13.3 (Equation 13.5) can also be used to approximate the transient storage term:

$$(mc_P)_{ii} \cdot \frac{dT_{ii}}{dt} \approx (mc_P)_{ii} \cdot \frac{T_{ii} - T_{ii}^{t-\Delta t}}{\Delta t} \qquad (17.14)$$

where Δt is the simulation timestep (some BPS tools use a shorter time interval).

The heat loss term is calculated using the method described by Equation 17.2:

$$q_{loss,ii} = (UA)_{ii} \cdot (T_{ii} - T_z) \qquad (17.15)$$

where $(UA)_{ii}$ is a heat loss coefficient (W/K) for the stratum, an input parameter required from the user.

Instances of Equations 17.12 to 17.15 are formed for each stratum. Any stratum including inlet and outlet ports will consider these mass flows as well. When solved in conjunction with Equation 17.11 for given inlet conditions, this leads to a solution of the temperature distribution within the tank, as well as the outlet conditions.

Most applications of this type of stratified tank model include methods for dealing with mixing caused by buoyancy. This can occur when, for example, the water entering the tank at state 1 is colder than T_i. If the result of the solution of the energy balances is such that $T_i < T_{ii}$, then the model assumes that buoyancy forces will cause these two strata to instantaneously mix to a common temperature equal to their average.

Many variants of this basic stratified tank modelling approach are employed in BPS tools. Some support the inclusion of immersed heat exchangers and immersed heating elements, and some include additional terms in the stratum energy balance of Equation 17.12 to account for heat transfer in the vertical direction caused by conduction through the tank's solid materials.

17.6 REQUIRED READING

Reading 17–A

Zhou *et al.* (2014) compares and contrasts the HVAC modelling approaches of three BPS tools.

Read this article in its entirety and find answers to the following questions:

1. Models of which HVAC components are examined?

2. Describe the main differences in the way the three BPS tools simulate the performance of HVAC systems.

3. Do these three BPS tools employ transient or steady-state models to represent boilers, chillers, and cooling towers?

4. How does the structure of the boiler models used by the three BPS tools compare with the approaches described in Section 17.2?

5. Do the chiller models of the three BPS tools use the same independent variables? What are they? How do these compare with the heat pump (also a vapour-compression refrigeration cycle) models described in Section 17.3?

6. Why are some differences observed in the predictions of the chiller models of the two BPS tools examined? (See Figure 13 in the article.)

17.7 SOURCES FOR FURTHER LEARNING

- Glembin *et al.* (2013) review existing boiler models that have been developed for BPS and propose a new model that considers transient effects.

- Blervaque *et al.* (2016) contrast steady-state empirical, transient empirical, and detailed thermodynamic models for variable-speed air-to-air heat pumps and compare their predictions to measurements.

- Allard *et al.* (2011) compare five models for stratified tanks and contrast their predictions to measured data.

17.8 SIMULATION EXERCISES

Exercise 17–A

Revert your BPS tool inputs to represent once again the case of Exercise 16–E. In that case you explicitly represented a hydronic heating system, but the radiator was supplied with water from a heat source at a constant temperature. Now you will replace that heat source with an explicit model of a boiler.

Consult your BPS tool's help file or technical documentation to determine its options for explicitly modelling boilers. Does it support methods like those outlined in Section 17.2? Describe the available methods.

Choose one of your tool's available methods and configure it to represent a small boiler (nominal heating capacity <10 kW). What input data are required? How does the chosen model calculate the boiler's efficiency? Consult technical information from manufacturers of boiler products that are available in your region to establish appropriate input values. Can you find sufficient data to establish all of the model's required input parameters? Make any necessary assumptions regarding the control of the boiler and the boiler model's parameters, such as the boiler's mass, stray heat losses, and efficiency.

Choose an appropriate timestep and conduct an annual simulation. What impact does this change have upon the annual space heating load? And upon the annual space cooling load?

Examine the simulation predictions for March 6. How does the boiler's fuel consumption and efficiency vary over the day? And the temperature of the water supplied to the radiator? Are these results consistent with your expectations?

Exercise 17–B

You will now replace the boiler added in Exercise 17–A with a heat pump.

Consult your BPS tool's help file or technical documentation to determine its options for explicitly modelling heat pumps. Does it support methods like those outlined in Section 17.3? Describe the available methods.

Choose one of your tool's available methods and configure it to represent a small air-source or water-source heat pump (nominal heating capacity <10 kW). What input data are required? How does the chosen model calculate the heat pump's rate of electricity consumption? Consult technical information from manufacturers of heat pump products that are available in your region to establish appropriate input values. Can you find sufficient data to establish all of the model's required input parameters? Make any necessary assumptions regarding the control of the heat pump and the heat pump model's parameters, such as the impact of operating conditions on its heat output and electricity consumption.

Choose an appropriate timestep and conduct an annual simulation. What impact does this change have upon the annual space heating load? And upon the annual space cooling load?

Examine the simulation predictions for March 6. How does the heat pump's electricity consumption vary over the day? And the temperature of the water supplied to the radiator? Are these results consistent with your expectations?

17.9 CLOSING REMARKS

This chapter described some commonly used methods for treating HVAC energy conversion and storage components that can be used to explicitly represent their performance and predict their fuel and energy consumption. By focusing on a few components— combustion boilers, heat pumps, solar thermal collectors, and water storage tanks—it demonstrated that models for energy conversion and storage rely heavily upon empirical information. It was seen that greater levels of detail invariably demand more input parameters from the user.

Through the required reading and the simulation exercises you became aware of the approaches supported by your chosen BPS tool, and learned how to configure them.

This completes not only our treatment of the modelling of HVAC systems, but also the treatment of all of the heat and mass transfer processes covered by this book. In the next part of the book you will integrate all of the learning and skills you have acquired to simulate the performance of an actual building.

VI

Finale

CHAPTER 18

Culminating trial

T HROUGH the previous chapters you have gained an understanding of the methods used to simulate individual heat and mass transfer processes and you have developed skills at applying your chosen BPS tool. Now it is time to put all of this together. This chapter describes the book's final simulation exercises which integrate the knowledge and skills you have gained through the application of BPS to predict the thermal performance of an actual building.

Chapter learning objectives

1. Integrate and concretize the knowledge gained in the previous chapters.
2. Learn strategies for representing real buildings in your chosen BPS tool.
3. Practice allocating time resources in a simulation project.
4. Develop experience comparing simulation predictions to measurements and diagnosing causes of disagreements.
5. Acquire experience at selecting appropriate sources for gathering input parameters, and learning how to manage uncertain inputs.

18.1 OVERVIEW

This is called the *Culminating Trial* because it tests your ability to apply the knowledge and skills you have gained through the previous chapters. All the exercises up to this point have been based on a simple building in the form of a rectangular box with a single window. The simplicity of the *Base Case* allowed us to focus on single heat or mass transfer processes or individual aspects of HVAC systems with each chapter's variations.

You will have to manage much greater complexity in this *Culminating Trial*. You will have to consider all of the heat and mass transfer processes treated in preceding chapters, as well as the performance of the HVAC system. You will have to decide which aspects have the greatest impact *in this case*. Which should receive your greatest attention? Which of your chosen BPS tool's default models can you rely upon? When and where is it important to invoke optional capabilities?

And you will have to decide how to thermally zone the building. How many thermal zones should you use? How should heat and mass transfer between the zones be treated? Which geometric features should you include? Which have a significant impact on the building's thermal performance, and which are less important?

BPS users never operate with complete certainty over input data. You will face this in the *Culminating Trial*. For example, you are not given the emissivity of building materials in the longwave spectrum, because these have not been measured. What values should you assume? You will have to search for appropriate sources of data. In some cases you can base your choices on your experiences with the previous simulation exercises. But in other cases, you might have to conduct sensitivity studies to examine how much of an impact these uncertain inputs might have upon your simulation predictions.

Time is always a finite resource. You will have to decide how to best allocate your time. Should you invest more time on defining the building's geometry, or on explicitly representing the HVAC system and its control? Which will have the greatest impact upon the accuracy of your simulation predictions?

The following sections describe the details of the *Culminating Trial* and the specific objectives of the exercises. But before you begin, you should revisit Reading 1–A, paying particular attention to Section 7.

18.2 APPROACH

The following sections describe the building and its HVAC systems in sufficient detail to enable you to prepare a representation in your chosen BPS tool. This is followed by a series of simulation exercises. These exercises should be conducted sequentially as they build upon each other.

You will see numerous references to the book's companion website, which contains additional information, such as photos, drawings, weather data, and other information that you will require to prepare your BPS representation and for conducting simulations. The book's companion website also contains measured data with uncertainty estimates against which you will compare your simulation predictions. To maximize the learning value of the exercises it is important that you not look at these measured data until instructed to do so.

You will be acquiring many data and making numerous assumptions and decisions as you progress through the simulation exercises. You should create a written report to document these choices. This is always good practice when creating a BPS representation, as this information can quickly become blurred during the process. The instructions for each simulation exercise provide some guidance on what kinds of information to add to your report.

All times mentioned in the *Culminating Trial* description, data files, weather files, and measured results are recorded in *standard time*, specifically in the Eastern Time Zone.

18.3 OBJECTIVES OF ANALYSES

The majority of the simulation exercises pertain to a five-day period in February 2019. You will be examining a number of simulation predictions from this period, such as the rate of solar radiation transmitted through the windows and air infiltration rates. And you will compare your simulation predictions against measured data over one particular day. This will include temporal predictions of the solar irradiance to an external surface, zone air temperatures, and the rate of heat addition/extraction to HVAC terminal devices.

You should keep these objectives in mind as you devise strategies for representing the *Culminating Trial* in your chosen BPS tool.

18.4   CASE STUDY BUILDING

An unoccupied research house located on the Carleton University campus in Ottawa is used as the object of the *Culminating Trial*. A photo of this facility as viewed from the southwest can be seen in Figure 1.1. You can see additional photos on the book's companion website. The house is located at 45.39 °N latitude and 75.70 °W longitude, at an elevation of 63 m above sea level. There are no surrounding trees, large objects, or buildings that cast shadows on the house.

The heavily glazed façade seen in Figure 1.1 faces slightly east of due south, with an azimuth angle of 10° (refer to Figure 9.3). Although not facing in a cardinal direction, it is identified as the *south* façade. Likewise, the other façades are identified as *east*, *north*, and *west*.

The house has a wood-frame construction with a footprint of 6.1 m by 12.2 m and includes a full-height basement and two above-grade storeys.

18.5   BUILDING ENVELOPE

The above-grade walls on the east, south, and west façades have a nominal U-value of 0.21 W/m² K while that on the north façade has a nominal U-value of 0.12 W/m² K. Details of the materials, dimensions, and fasteners of these envelope assemblies are provided on the book's companion website.

The insulation added to the attic has a nominal U-value of 0.11 W/m² K. The insulation under the basement floor and around the foundation's structural footings has a nominal U-value of 0.36 W/m² K while the insulation added to the basement walls has a nominal U-value of 0.21 W/m² K. Again, details on these assemblies are provided on the book's companion website.

The insulated glazing units of all windows are of the same construction. They are triple-glazed and filled with argon. The outer layer is a 3 mm thick sheet of clear glass. The inner two layers are also 3 mm thick but these contain low-emissivity coatings on surfaces 3 and 5, as illustrated in Figure 1.3. The thickness of the argon gas layers is 13 mm.

The thermophysical and radiation properties of the glazings are identical to those of the *Base Case* and are provided in Tables 1.6 and 1.7. The insulated glazing units are supported by frames fabricated from extruded polyvinyl.

Most of the house's windows are placed on the south façade. The east and west façade each include a modest amount of window area, whereas no windows face north. The dimensions and locations of each of the house's windows are provided in drawings that can be found on the book's companion website. Except when noted, all windows remained closed for the simulation exercises of this *Culminating Trial* and the roller blinds that can be seen in Figure 1.1 were fully retracted.

The building's airtightness was assessed through a depressurization test using a blower door apparatus, as described in Section 15.3. A regression of these measured data to the power law relationship of Equation 15.4 produced a best fit with the coefficients $C = 0.01689 \, \text{m}^3/\text{s Pa}^n$ and $n = 0.68$. This corresponds approximately to 1.3 ac/h at 50 Pa depressurization.

18.6 HVAC

The research house possesses redundant energy conversion, storage, and distribution systems to enable research on numerous topics with minimal switch-over time required. There is a large array of evacuated-tube solar thermal collectors (seen in Figure 1.1), two buried seasonal thermal stores, an air-source heat pump coupled to a rock-bed thermal store, a water-to-water heat pump, radiant hydronic floors, supply and return ducts, an energy recovery ventilator connected to dedicated ducting, and an auxiliary electric boiler.

Much of the above-mentioned HVAC equipment was inoperative during the February 2019 period, and therefore can be safely ignored for many of the simulation exercises. During this time the house was heated and cooled by the hydronic radiant floors that fully cover the ground storey and second storey. The hydronic floors were supplied by a loop containing a circulation pump, hot and cold water tanks (stores), and a 12 kW modulating electric boiler. This system configuration is represented schematically in Figure 18.1.

A water-to-water heat pump was coupled to the hot and cold stores and was used to cool the cold store while concurrently heating the hot store. The pump (indicated by ▶) that circulates water between the cold store and the heat pump's evaporator heat exchanger and the pump that circulates water between the hot store and the heat pump's condenser heat exchanger can be seen in the figure.

You can find performance characteristics of this heat pump in Brideau *et al.* (2016a), who measured its performance and proposed a transient empirical model suitable for use in BPS. Controls were configured to activate the heat pump and the evaporator and condenser pumps to maintain the cold store between 8 °C and 10 °C.

Air sensors located on each storey sensed the average air temperature on that storey. If either storey called for heating (or cooling), then the circulation pump indicated in the figure would cycle on. Due to the arrangement of the three-way fixed-position diverting (indicated by ○) and converging (indicated by ●) valves, this would circulate water to the hydronic floors on both storeys of the house concurrently. If the demand was for heating, then the three-way controlled valve (indicated by ■) would be actuated to draw water from the hot store. And if the demand was for cooling, then the three-way controlled valve would be actuated to draw water from the cold store.

A storey called for heating when its air temperature dropped below 19.5 °C and continued to call for heating until this temperature rose above 20.5 °C. A storey called for cooling when its air temperature rose above 23.5 °C and continued to call for cooling until this temperature dropped below 22.5 °C.

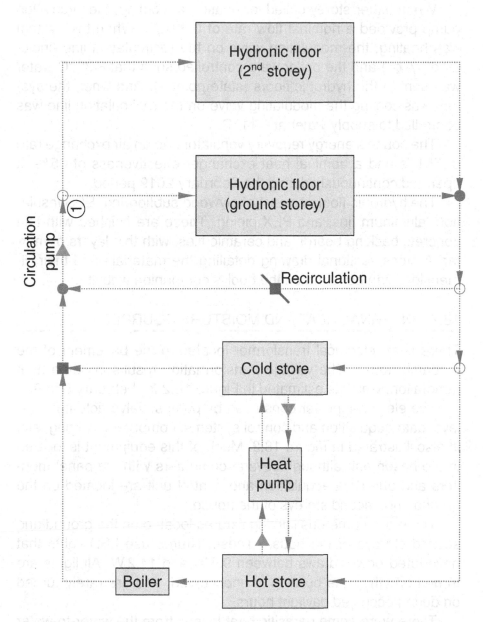

Figure 18.1: HVAC system configuration (minus energy recovery ventilator) for February 2019 period (pumps identified by ▶, controlled valves by ■, fixed-position diverting valves by ○, and fixed-position converging valves by ●)

When either storey called for heating or cooling, the circulation pump provided a nominal flow rate of 0.38 L/s. When the system was heating, the modulating valve on the recirculation line (indicated by ◢) and the boiler were controlled such that ~30 °C water was sent to the hydronic floors (state point 1). And when the system was cooling the modulating valve on the recirculation line was controlled to supply water at ~11 °C.

The house's energy recovery ventilator had an air exchange rate of 55 L/s and a nominal heat exchanger effectiveness of 65%. It operated continuously during the February 2019 period.

The hydronic floors consist of plywood subflooring, EPS insulation, aluminum fins, and PEX piping. These are finished with thin concrete backing boards and ceramic tiles, with thin layers of mortar. A cross-sectional drawing detailing the materials and their dimensions can be found on the book's companion website.

18.7   INTERNAL HEAT AND MOISTURE SOURCES

There is an electrical transformer located in the basement of the research house. The voltage transformation results in some heat generation, which is estimated in Figure 18.2 for February 5 to 9.

The electrical power consumed by pumps, valve actuators, relays, data acquisition and control systems, computers, lighting, etc. is also illustrated in Figure 18.2. Much of this equipment is located in the basement, although several computers with flat-panel monitors and one data acquisition and control unit are located on the ground and second storeys of the house.

There are numerous lighting fixtures located on the ground and second storeys of the house. These fixtures use LED lights that have rated power draws between 9.5 W and 14.2 W. All lights are turned off when the building is unoccupied, and are rarely turned on during occupied daylight hours.

There were some parasitic heat losses from the water-to-water heat pump (located in the basement) that was running during the February period. As well, there were some heat losses from water storage tanks located in the basement. These heat losses are also estimated in Figure 18.2.

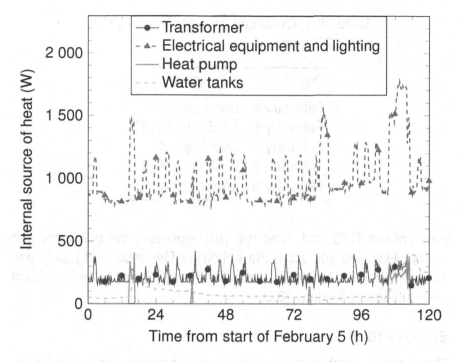

Figure 18.2: Non-occupant internal sources of heat for the February 2019 period (corresponding data file available on book's companion website)

The building was unoccupied except for the times indicated in Table 18.1. During these times the occupants (typically one to three people) were doing light work, such as working with a laptop computer, taking measurements using a portable data acquisition device, or working on configuring experimental equipment.

18.8 SIMULATION EXERCISES

Exercise 18–A

The book's companion website includes a number of photographs of the research house as well as dimensioned drawings in plan and elevation views. Study this information and then review the material on zoning that was presented in Chapter 2 (Section 2.2).

Establish an appropriate strategy for abstracting the building in

Table 18.1: Occupant presence during February 2019 period

Day	Occupancy
February 5	unoccupied
February 6	13h53 to 14h43
February 7	unoccupied
February 8	8h56 to 12h18
February 9	unoccupied

your chosen BPS tool. How will you represent the building's geometry? How will you zone the building? Describe and justify your approach in your report. Explain why it is appropriate for the objectives outlined in Section 18.3.

Exercise 18–B

There are a number of cross-sectional drawings of the building's opaque envelope assemblies included on the book's companion website. These specify materials, dimensions, and fastener details. After studying this information, add a table to your report listing all the thermophysical and radiation properties of opaque envelope materials that you will require to conduct your simulations. What sources of information can you use to acquire the necessary data? Consider sources such as handbooks, published manufacturer data, articles and papers, and BPS tool databases.

For each item in your table, indicate the range of possible values based upon the data you have found. Which parameters are uncertain? Select the values to use in your simulations and justify them. Based upon the experience you accumulated during the exercises of the previous chapters, which of the estimated parameters do you think will contribute the greatest uncertainty in your simulation predictions?

Review the material on the temperature dependence of the effective thermal conductivity of insulation materials presented in Chapter 13 (Section 13.2 and Reading 13–B). Have you considered the temperature dependence of the effective thermal conductivity of any of the insulation materials? If so, what is the source of the data you used. If not, then justify your approach.

Exercise 18–C

Review the material on thermal bridging that was presented in Chapter 13 (Section 13.7). Based on this and the cross-sectional drawings you reviewed during Exercise 18–B decide how you will treat thermal bridges.

Add a section to your report to explain and justify your approach. Explain why it is appropriate for the task at hand.

Exercise 18–D

The dimensions and locations of each of the house's windows are provided in drawings that can be found on the book's companion website. The dimensions of the window frames and the transparent portion of the insulated glazing units can be determined from these drawings.

How will you treat the influence of window spacers and frames? Will you represent each window as a separate entity in your BPS tool, or will you combine some? Add a section to your report to explain and justify your approach.

Exercise 18–E

What approach will you use for representing the HVAC system? Will you employ an idealized approach, or explicitly represent the HVAC components?

As indicated in Section 18.3, you are required to predict the rate of heat addition/extraction to the hydronic floors but not the electrical energy consumption of the heat pump and the pumps. Will you explicitly represent all the energy conversion and storage components shown in Figure 18.1, or just the components of the energy distribution system?

How will you represent heat transfer between the hydronic floors and the building's interior environment? Will you consider its transient response to changes in the circulating pump's operation? How will you represent the control behaviour described in Section 18.6?

Add a section to your report to explain and justify your approach.

Exercise 18–F

Devise appropriate strategies for treating the following topics:

- Solar shading, diffuse solar irradiance from the sky, ground-reflected solar irradiance, and the distribution of solar irradiance to internal surfaces.
- Convection heat transfer at internal and external surfaces.
- View factors for longwave radiation exchange at internal and external surfaces, and the sky temperature.
- Internal heat and moisture sources.
- Air infiltration and transfer airflow between zones.
- Heat transfer to the ground.

Which of your chosen BPS tool's default and optional modelling methods will you employ? Base your decisions upon the theory presented in the book, the required readings, and the experience you accumulated during the exercises of the previous chapters.

Add a short description to your report to explain and justify the strategies you have chosen for each of the above topics. Which of your choices do you think will have the greatest impact upon simulation predictions?

Exercise 18–G

An EPW formatted file with weather data that were measured at the site of the research house can be found on the book's companion website.

Download this file and examine the weather conditions for February 5 to 9 using your BPS tool or one of the weather utilities suggested in Exercise 8–C. You may also want to examine supplementary weather information, such as precipitation and depth of the snow cover that is also posted on the book's companion website.

How to you think the building will perform under these weather conditions? How do you expect infiltration rates to vary over this period? Which days during this period will require the greatest heat addition from the HVAC system? Do you think the HVAC system will have to provide any cooling? Estimate the peak rate of solar transmission through the windows based upon hand calculations and some simplifying approximations.

Describe your expectations for the building's performance in your report.

Exercise 18–H

Execute the decisions you took and input the data you gathered during Exercise 18–A to Exercise 18–F to create a representation of the research house using your selected BPS tool.

Conduct sensitivity studies to explore the impact of the modelling strategies and input parameters that you predicted would be the most impactful. Hone your modelling approaches and input data as appropriate.

Now perform a simulation and appraise the predictions for February 5 through 9. Scrutinize your simulation predictions. Examine, for example, zone air temperatures, the rate of solar radiation transmitted through the windows, internal heat gains, air infiltration rates, and the rate of HVAC heat injection/extraction.

Are the simulation predictions consistent with the expectations you formed in Exercise 18–G? If not, think of some possible explanations and explore these possible causes.

Don't assume that your BPS representation is valid or accurate just because your tool does not issue any error/warning messages. It is important that you scrutinize your results in detail so that you develop trust in your predictions.

Record your observations in your report.

Exercise 18–I

Refine your BPS representation of the research house based upon what you learned during Exercise 18–H and then perform another simulation. We will focus on the results for February 9, a cool and sunny day.

Create a heat transfer versus time graph and plot your simulation predictions for the solar irradiance (W/m^2) incident upon the exterior surface of the building's south façade. Compare this graph to the measured results that can be found on the book's companion website.

Are your simulation predictions in agreement with the measured data (within the measurement uncertainty estimates)? Are they temporally aligned? Are the peak values similar?

If your simulation predictions are not in good agreement with the measurements, then postulate possible causes. Think in particular about the solar geometry and ground-reflected irradiance topics that were treated in Chapter 9. Investigate these possibilities and effect changes to your input parameters that you can justify. Record the details of what you have changed and the reasons why in your report.

Exercise 18–J

Refine your BPS representation of the research house based upon what you learned during Exercise 18–I and then perform another simulation.

Create a temperature versus time graph for February 9 and plot your simulation predictions for the zone air temperature for the building's ground storey. Compare this graph to the measured results that can be found on the book's companion website.

Are your simulation predictions in agreement with the measured data (within the measurement uncertainty estimates)? Are they temporally aligned? Are the peak values similar?

If your simulation predictions are not in good agreement with the measurements, then postulate possible causes. Think in particular about the zone energy and internal surface energy balances that were treated in Chapter 2. The treatment of which terms appearing in these energy balances might be leading to the differences you are observing? Investigate these possibilities and effect changes to your input parameters that you can justify. Record the details of what you have changed and the reasons why in your report.

Exercise 18–K

Refine your BPS representation of the research house based upon what you learned during Exercise 18–J and then perform another simulation.

Create a temperature versus time graph for February 9 and plot your simulation predictions for the zone air temperature for the building's basement. Compare this graph to the measured results that can be found on the book's companion website.

Are your simulation predictions in agreement with the measured data (within the measurement uncertainty estimates)? Are they temporally aligned? Are the peak values similar?

If your simulation predictions are not in good agreement with the measurements, then postulate possible causes. Think in particular about the treatment of heat transfer to the ground that was treated in Chapter 12. Investigate these possibilities and effect changes to your input parameters that you can justify. Record the details of what you have changed and the reasons why in your report.

Exercise 18–L

Refine your BPS representation of the research house based upon what you learned during Exercise 18–K and then perform another simulation.

Create a heat transfer versus time graph for February 9. Plot the combined rate of heat injection or extraction to the hydronic floors in both the ground storey and the second storey versus time. Plot the values in kW where positive values indicate a heat injection from the HVAC system to the hydronic floors, and negative values indicate a heat extraction by the HVAC system. Compare this graph to the measured results that can be found on the book's companion website.

Are your simulation predictions in agreement with the measured data (within the measurement uncertainty estimates)? Does the simulation predict the correct direction of heat transfer during the nighttime and during the daytime? Are the HVAC system operating times correctly predicted?

If your simulation predictions are not in good agreement with the measurements, then postulate possible causes. Think in particular about the treatment of the hydronic floors and the HVAC controls that were discussed in Chapter 16. Are there transient effects that are not well represented in your simulation? Consider as well the treatment of convection (Chapter 4) and longwave radiation (Chapter 5) from the hydronic floors. Investigate these possibilities and effect changes to your input parameters that you can justify. Record the details of what you have changed and the reasons why in your report.

Exercise 18–M

Refine your BPS representation of the research house based upon what you learned during Exercise 18–L.

Choose one of the other trial periods described on the book's companion website. The operational characteristics of the research house and HVAC system vary for each one of these trial periods. Carefully review the descriptions of how the house was operated (window openings, window blinds, internal sources of heat, HVAC, etc.) for your chosen trial period and make any necessary adjustments to the inputs of your BPS tool. Download the corresponding weather data and conduct a simulation.

The measured performance data for each trial period are also posted on the book's companion website. Compare your simulation predictions to these results.

Are your simulation predictions in agreement with the measured data (within the measurement uncertainty estimates)? If not, revisit some of the decisions you took in Exercise 18–A to Exercise 18–F. Conduct additional simulations to test the impact of some of these decisions.

In your report list the five aspects that you think contribute the greatest uncertainty in predicting the required results. What have you learned from this exercise? What will you do differently on your next simulation project? How would you change the way you allocate time to tasks, or prioritize tasks?

18.9 CLOSING REMARKS

Through this chapter's simulation exercises you had to integrate the knowledge and skills you had gained during previous chapters to apply your chosen BPS tool to predict the thermal performance of an actual building. You had to deal with many challenging topics, such as zoning the building, deciding which geometrical features to represent, and choosing which tool default and optional modelling approaches to employ. And, importantly, you had to make choices about how to best allocate your time to achieve the objectives of the exercises.

There is no single *right* way to deal with these complex topics. By comparing your simulation predictions to the measured results, you doubtlessly learned something about the appropriateness of some of your approaches and choices. And this new experience will hopefully inform your approaches in the future.

CHAPTER 19

Next steps

THIS brings our tour through BPS to an end. By successfully completing all of the book's learning elements, you now have a solid understanding of the fundamentals. You can appreciate the inherent (and necessary) simplifications in some models and you realize which models are applicable in which situations. You understand the implications of modelling choices and default modelling methods and input data, and you realize which modelling choices and input data have the greatest impact upon simulation predictions.

You have developed a certain proficiency at applying your chosen BPS tool through conducting the simulation exercises and the *Culminating Trial*. But don't stop there. Learn how to use other BPS tools, perhaps repeating some or all of the book's simulation exercises. Don't be limited to a single tool. Develop the expertise to employ multiple tools, because they all have strengths and weaknesses and none meets the needs of every situation.

For your future work, don't become complacent about accepting tool default models and data. Use the knowledge you have accumulated through studying this book to critically examine BPS tools and to determine how to best apply them. Go beyond the documentation and training that shows you how to operate the tools. Become knowledgeable about the models they employ and assess them critically. Become comfortable consulting technical documentation, journal articles, conference papers, and theses that document their models. In some cases you will find it necessary to delve into source code to really understand what is going on.

This book has emphasized depth at the expense of breadth. Its scope was necessarily limited to heat and mass transfer processes relevant to the building's form and fabric and HVAC systems. Moreover, space and time precluded an in-depth treatment of important topics such as model abstraction. There is, of course, much more to learn. You could continue with a study of topics such as daylighting, occupant comfort, acoustics, electrical energy conversion and storage systems, managing uncertainty, etc.

There are many sources you can turn to for further learning. The International Building Performance Simulation Association (IBPSA) provides a wealth of information. Consult its website, read its newsletter, and peruse the proceedings of its conferences. Join one of its chapters (many offer free membership) and consider attending its biennial international conference, or a conference organized by one of its regional chapters. These are great venues to meet practitioners and researchers working in the BPS field. A number of the required readings were taken from IBPSA's *Journal of Building Performance Simulation*. Its repository includes thousands of pages of articles on original research related to BPS.

Understanding the fundamentals—the focus of this book—is critical. But, of course, this is not sufficient. It is also necessary for BPS users to develop the necessary skills for collaborating and interacting with building designers. Skills such as interpreting design questions and translating them into simulation analyses, interpreting results, and providing timely and appropriate feedback to inform design teams must also be cultivated. Try to guide your life-long learning to acquire these skills. Just don't stop learning.

References

Alamdari F and Hammond G (1983), Improved data correlations for buoyancy-driven convection in rooms, *Building Services Engineering Research and Technology*, 4(3):106–112, URL http://dx.doi.org/10.1177%2F014362448300400304.

Allard Y, Kummert M, Bernier M, and Moreau A (2011), Intermodal comparison and experimental validation of electrical water heater models in TRNSYS, in *Proceedings of Building Simulation 2011*, International Building Performance Simulation Association, Sydney, Australia, URL http://www.ibpsa.org/proceedings/BS2011/P_1310.pdf.

ASHRAE (2016), ASHRAE handbook of heating, ventilating, and air-conditioning systems and equipment, Technical report, American Society for Heating, Refrigeration, and Air-Conditioning Engineers, Atlanta, USA.

ASHRAE (2017), ASHRAE handbook of fundamentals, Technical report, American Society for Heating, Refrigeration, and Air-Conditioning Engineers, Atlanta, USA.

ASTM (2012), Standard G173-03: Standard tables for reference solar spectral irradiances: Direct normal and hemispherical on 37° tilted surface, Technical report, American Society for Testing and Materials, West Conshohocken, USA, URL http://dx.doi.org/10.1520/G0173-03R12.

ASTM (2014), Standard E490-00a: Standard solar constant and zero air mass solar spectral irradiance tables, Technical report, American Society for Testing and Materials, West Conshohocken, USA, URL http://dx.doi.org/10.1520/E0490-00AR14.

Awbi H and Hatton A (1999), Natural convection from heated room surfaces, *Energy and Buildings*, 30(3):233–244, URL `http://dx.doi.org/10.1016/S0378-7788(99)00004-3`.

Axley J (2007), Multizone airflow modeling in buildings: History and theory, *HVAC and R Research*, 13(6):907–928, URL `http://dx.doi.org/10.1080/10789669.2007.10391462`.

Ayres J and Stamper E (1995), Historical development of building energy calculations, *ASHRAE Transactions*, 101(1):841–848, URL `https://www.techstreet.com/standards/ch-95-09-1-historical-development-of-building-energy-calculations?product_id=1716390`.

Beausoleil-Morrison I (2002), The adaptive simulation of convective heat transfer at internal building surfaces, *Building and Environment*, 37(8-9):791–806, URL `http://dx.doi.org/10.1016/S0360-1323(02)00042-2`.

Beausoleil-Morrison I (2019), Learning the fundamentals of building performance simulation through an experiential teaching approach, *Journal of Building Performance Simulation*, 12(3):308–325, URL `http://dx.doi.org/10.1080/19401493.2018.1479773`.

Belcher S, Hacker J, and Powell D (2005), Constructing design weather data for future climates, *Building Services Engineering Research and Technology*, 26(1):49–61, URL `http://dx.doi.org/10.1191/0143624405bt112oa`.

Bennet IE and O'Brien W (2017), Office building plug and light loads: Comparison of a multi-tenant office tower to conventional assumptions, *Energy and Buildings*, 153:461–475, URL `http://dx.doi.org/10.1016/j.enbuild.2017.08.050`.

Berardi U (2019), The impact of aging and environmental conditions on the effective thermal conductivity of several foam materials, *Energy*, 182:777–794, URL `http://dx.doi.org/10.1016/j.energy.2019.06.022`.

Berardi U and Naldi M (2017), The impact of the temperature dependent thermal conductivity of insulating materials on the effective building envelope performance, *Energy and Buildings*, 144:262–275, URL http://dx.doi.org/10.1016/j.enbuild.2017.03.052.

Blervaque H, Stabat P, Filfli S, Schumann M, and Marchio D (2016), Variable-speed air-to-air heat pump modelling approaches for building energy simulation and comparison with experimental data, *Journal of Building Performance Simulation*, 9(2):210–225, URL http://dx.doi.org/10.1080/19401493.2015.1030862.

Brideau S, Beausoleil-Morrison I, and Kummert M (2016a), Empirical model of a 11 kW (nominal cooling) R134a water-water heat pump, in *Proc. eSim 2016*, Hamilton, Canada, URL http://www.ibpsa.org/proceedings/eSimPapers/2016/20-118-eSim2016.pdf.

Brideau S, Beausoleil-Morrison I, Kummert M, and Wills A (2016b), Inter-model comparison of embedded-tube radiant floor models in BPS tools, *Journal of Building Performance Simulation*, 9(2):190–209, URL http://dx.doi.org/10.1080/19401493.2015.1027065.

Bueno B, Norford L, Hidalgo J, and Pigeon G (2013), The urban weather generator, *Journal of Building Performance Simulation*, 6(4):269–281, URL http://dx.doi.org/10.1080/19401493.2012.718797.

Cascone Y, Corrado V, and Serra V (2011), Calculation procedure of the shading factor under complex boundary conditions, *Solar Energy*, 85(10):2524–2539, URL http://dx.doi.org/10.1016/j.solener.2011.07.011.

Čekon M (2015), Accuracy analysis of longwave sky radiation models in the MZELWE module of the ESP-r program, *Energy and Buildings*, 103:147–158, URL http://dx.doi.org/10.1016/j.enbuild.2015.06.039.

Ceylan HT and Myers GE (1980), Long-time solutions to heat-conduction transients with time-dependent inputs, *Journal of Heat Transfer*, 102(1):115–120, URL http://dx.doi.org/1 0.1115/1.3244221.

Chan YC and Tzempelikos A (2013), Analysis and comparison of absorbed solar radiation distribution models in perimeter building zones, *ASHRAE Transactions*, 119(2):129–145, URL https://www.techstreet.com/standards/de-13-012-ana lysis-and-comparison-of-absorbed-soalr-radiation-d istribution-models-in-perimeter-building-zones?pro duct_id=1867169.

Chatziangelidis K and Bouris D (2009), Calculation of the distribution of incoming solar radiation in enclosures, *Applied Thermal Engineering*, 29(5–6):1096–1105, URL http://dx.doi.org /10.1016/j.applthermaleng.2008.05.026.

Chen D (2015), Three-dimensional steady-state ground heat transfer for multi-zone buildings, *Journal of Building Performance Simulation*, 8(2):44–56, URL http://dx.doi.org/10. 1080/19401493.2013.866696.

Chen Q (2009), Ventilation performance prediction for buildings: A method overview and recent applications, *Building and Environment*, 44(4):848–858, URL http://dx.doi.org/http: //dx.doi.org/10.1016/j.buildenv.2008.05.025.

Churchill SW and Chu HH (1975), Correlating equations for laminar and turbulent free convection from a vertical plate, *International Journal of Heat and Mass Transfer*, 18(11):1323–1329, URL http://dx.doi.org/https://doi.org/10.1016/0017 -9310(75)90243-4.

CIBSE (2015), CIBSE Guide A: Environmental design, Technical report, The Chartered Institution of Building Services Engineers, London, UK.

Clark G and Allen C (1978), The estimation of atmospheric radiation for clear and cloudy skies, in *Proceedings of the 2nd Na-*

tional Passive Solar Conference, volume II, pages 675–678, American Solar Energy Society, Philadelphia, USA.

Clarke J (2001), *Energy Simulation in Building Design*, Butterworth-Heinemann, Oxford, UK, 2nd edition, URL https: //www.sciencedirect.com/book/9780750650823/energy-simulation-in-building-design.

Clarke J (2015), A vision for building performance simulation: a position paper prepared on behalf of the IBPSA board, *Journal of Building Performance Simulation*, 8(2):39–43, URL http: //dx.doi.org/10.1080/19401493.2015.1007699.

Clarke J and Hensen J (2015), Integrated building performance simulation: Progress, prospects, and requirements, *Building and Environment*, 91:294–306, URL http://dx.doi.org/1 0.1016/j.buildenv.2015.04.002.

Cook MJ, Ji Y, and Hunt GR (2003), CFD modelling of natural ventilation: Combined wind and buoyancy forces, *International Journal of Ventilation*, 1(3):169–179, URL http://dx.doi.o rg/10.1080/14733315.2003.11683632.

Crawley DB and Barnaby CS (2019), *Building Performance Simulation for Design and Operation*, chapter Weather and climate in building performance simulation, pages 191–220, Routledge, Abingdon, UK, 2nd edition, URL https://www.rout ledge.com/Building-Performance-Simulation-for-Desi gn-and-Operation/Hensen-Lamberts/p/book/9781138392 199.

Crawley DB, Hand J, and Lawrie LK (1999), Improving the weather information available to simulation programs, in *Proceedings of Building Simulation 1999*, International Building Performance Simulation Association, Kyoto, Japan, URL http://www.ibps a.org/proceedings/BS1999/BS99_P-03.pdf.

Crawley DB, Hand JW, Kummert M, and Griffith BT (2008), Contrasting the capabilities of building energy performance simulation programs, *Building and Environment*, 43(4):661–673, URL http://dx.doi.org/10.1016/j.buildenv.2006.10.027.

Crawley DB and Lawrie LK (2015), Rethinking the TMY: Is the typical meteorological year best for building performance simulation?, in *Proc. Building Simulation 2015*, pages 2655–2662, Hyderabad, India, URL http://www.ibpsa.org/proceeding s/BS2015/p2707.pdf.

Curcija C, Vidanovic S, Hart R, Jonsson J, and Mitchell R (2018), WINDOW technical documentation, Technical report, Lawrence Berkeley National Laboratory, Berkeley, USA, URL https://windows.lbl.gov/sites/default/files/Downlo ads/WINDOW%20Technical%20Documentation.pdf.

de Almeida Rocha AP, Oliveira RC, and Mendes N (2017), Experimental validation and comparison of direct solar shading calculations within building energy simulation tools: Polygon clipping and pixel counting techniques, *Solar Energy*, 158:462–473, URL http://dx.doi.org/10.1016/j.solener.2017. 10.011.

de Gastines M, Correa É, and Pattini A (2019), Heat transfer through window frames in EnergyPlus: model evaluation and improvement, *Advances in Building Energy Research*, 13(1):138–155, URL http://dx.doi.org/10.1080/17512 549.2017.1421098.

Delcroix B, Kummert M, Daoud A, and Hiller M (2013), Improved conduction transfer function coefficients generation in TRNSYS multizone building model, in *Proceedings of Building Simulation 2013*, International Building Performance Simulation Association, Chambéry, France, URL http://www.ibps a.org/proceedings/BS2013/p_1192.pdf.

Deru M, Field K, Studer D, Benne K, Griffith B, Torcellini P, Liu B, Halverson M, Winiarski D, Rosenberg M, Yazdanian M, Huang J, and Crawley D (2011), US Department of Energy commercial reference building models of the national building stock, Technical report, National Renewable Energy Laboratory, Golden, USA, URL https://www.nrel.gov/docs/fy11 osti/46861.pdf, NREL/TP-5500-46861.

Duffie JA and Beckman WA (2006), *Solar Engineering of Thermal Processes*, John Wiley & Sons, Hoboken, USA, 3rd edition.

Dumitrascu L and Beausoleil-Morrison I (2020), A model for predicting the solar reflectivity of the ground that considers the effects of accumulating and melting snow, *Journal of Building Performance Simulation*, 13(3):334–346, URL http://dx.doi.org/10.1080/19401493.2020.1728383.

Emmerich SJ (2006), Simulated performance of natural and hybrid ventilation systems in an office building, *HVAC&R Research*, 12(4):975–1004, URL http://dx.doi.org/10.1080/10789669.2006.10391447.

Evangelisti L, Guattari C, and Asdrubali F (2019), On the sky temperature models and their influence on buildings energy performance: A critical review, *Energy and Buildings*, 183:607–625, URL http://dx.doi.org/10.1016/j.enbuild.2018.11.037.

Fisher D and Pedersen C (1997), Convection heat transfer in building energy and thermal load calculations, *ASHRAE Transactions*, 103(2):137–148, URL https://pdfs.semanticscholar.org/3be3/0d071201b1623ee51b497bb331477b981fb1.pdf.

Francisco SC, Raimundo AM, Gaspar AR, Oliveira AVM, and Quintela DA (2014), Calculation of view factors for complex geometries using Stokes' theorem, *Journal of Building Performance Simulation*, 7(3):203–216, URL http://dx.doi.org/10.1080/19401493.2013.808266.

Furler R (1991), Angular dependence of optical properties of homogeneous glasses, *ASHRAE Transactions*, 97(2):1129–1133, URL https://escholarship.org/content/qt89j2v2tq/qt89j2v2tq.pdf.

GATC (1967), *Computer program for analysis of energy utilization in postal facilities: Volume 1 user's manual*, General American Transportation Corporation.

Glembin J, Rockendorf G, Betram E, and Steinweg J (2013), A new easy-to-parameterize boiler model for dynamic simulations, *ASHRAE Transactions*, 119(1):270–292, URL `https://www.techstreet.com/standards/da-13-024-a-new-eas y-to-parameterize-boiler-model-for-dynamic-simulat ions?product_id=1855958`.

Gunay HB, O'Brien W, and Beausoleil-Morrison I (2016), Implementation and comparison of existing occupant behaviour models in EnergyPlus, *Journal of Building Performance Simulation*, 6(6):567–588, URL `http://dx.doi.org/10.1080/194 01493.2015.1102969`.

Hensen JL (1991), *On the thermal interaction of building structure and heating and ventilating system*, Ph.D. thesis, Technische Universiteit Eindhoven, Eindhoven, The Netherlands, URL `ht tp://dx.doi.org/https://doi.org/10.6100/IR353263`.

Hensen JL and Lamberts R (2019), *Building Performance Simulation for Design and Operation*, chapter Building performance simulation—challenges and opportunities, pages 1–10, Routledge, Abingdon, UK, 2nd edition, URL `https://www.routle dge.com/Building-Performance-Simulation-for-Design -and-Operation/Hensen-Lamberts/p/book/978113839219 9`.

Hiller MD, Beckman WA, and Mitchell JW (2000), TRNSHD–a program for shading and insolation calculations, *Building and Environment*, 35(7):633–644, URL `http://dx.doi.org/10.10 16/S0360-1323(99)00051-7`.

Huang YJ (2019), Using satellite-derived solar radiation to create weather files of unprecedented accuracy and reliability, in *Proceedings of Building Simulation 2019*, pages 4846–4853, International Building Performance Simulation Association, Rome,

Italy, URL http://www.ibpsa.org/proceedings/BS2019/B
S2019_211389.pdf.

Incropera FP, DeWitt DP, Bergman TL, and Lavine AS (2007), *Fundamentals of Heat and Mass Transfer*, John Wiley & Sons, Hoboken, USA, 6th edition.

Iousef S, Montazeri H, Blocken B, and van Wesemael P (2019), Impact of exterior convective heat transfer coefficient models on the energy demand prediction of buildings with different geometry, *Building Simulation*, 12(5):797–816, URL http://dx.doi.org/10.1007/s12273-019-0531-7.

Johnson MH, Zhai ZJ, and Krarti M (2012), Performance evaluation of network airflow models for natural ventilation, *HVAC&R Research*, 18(3):349–365, URL https://www.tandfonline.com/doi/full/10.1080/10789669.2011.617291.

Jones NL, Greenberg DP, and Pratt KB (2012), Fast computer graphics techniques for calculating direct solar radiation on complex building surfaces, *Journal of Building Performance Simulation*, 5(5):300–312, URL http://dx.doi.org/10.1080/19401493.2011.582154.

Jorissen F, Reynders G, Baetens R, Picard D, Saelens D, and Helsen L (2018), Implementation and verification of the IDEAS building energy simulation library, *Journal of Building Performance Simulation*, 11(6):669–688, URL http://dx.doi.org/10.1080/19401493.2018.1428361.

Judkoff R and Neymark J (1995), *International Energy Agency Building Energy Simulation Test (BESTEST) and Diagnostic Method*, IEA/ECBCS Annex 21 Subtask C and IEA/SHC Task 12 Subtask B Report, URL https://www.osti.gov/biblio/90674-international-energy-agency-building-energy-simulation-test-bestest-diagnostic-method.

Judkoff R, Wortman D, O'Doherty B, and Burch J (1983), A methodology for validating building energy analysis simulations,

Technical Report TR-254-1508, Solar Energy Research Institute, Golden, USA, URL https://www.nrel.gov/docs/fy08 osti/42059.pdf.

Khalifa A and Marshall R (1990), Validation of heat transfer coefficients on interior building surfaces using a real-sized indoor test cell, *International Journal of Heat and Mass Transfer*, 33(10):2219–2236, URL http://dx.doi.org/10.1016/001 7-9310(90)90122-B.

Kramer SC, Gritzki R, Perschk A, Rösler M, and Felsmann C (2015), Fully parallel, OpenGL-based computation of obstructed area-to-area view factors, *Journal of Building Performance Simulation*, 8(4):266–281, URL http://dx.doi.org/10.10 80/19401493.2014.917700.

Kreith F and Kreider JF (1978), *Principles of Solar Engineering*, Hemisphere Publishing, Washington, USA.

Kruis N and Krarti M (2015), Kiva™: a numerical framework for improving foundation heat transfer calculations, *Journal of Building Performance Simulation*, 8(6):449–468, URL http://dx.d oi.org/10.1080/19401493.2014.988753.

Kruis N and Krarti M (2017), Three-dimensional accuracy with two-dimensional computation speed: using the Kiva™ numerical framework to improve foundation heat transfer calculations, *Journal of Building Performance Simulation*, 10(2):161–182, URL http://dx.doi.org/10.1080/19401493.2016.121 1177.

Kusuda T (1976), *NBSLD: The computer program for heating and cooling loads in buildings. NBS Building science series No. 69*, National Bureau of Standards, URL https://www.govinfo. gov/content/pkg/GOVPUB-C13-9bcc6856169c63cf2c5ab81 af189bd75/pdf/GOVPUB-C13-9bcc6856169c63cf2c5ab81af 189bd75.pdf.

Kusuda T (1999), Early history and future prospects of building system simulation, in *Proceedings of Building Simulation 1999*, pages 3–15, International Building Performance Simulation Association, Kyoto, Japan, URL http://www.ibpsa.org/proc eedings/BS1999/BS99_P-01.pdf.

Lauzet N, Rodler A, Musy M, Azam MH, Guernouti S, Mauree D, and Colinart T (2019), How building energy models take the local climate into account in an urban context – a review, *Renewable and Sustainable Energy Reviews*, 116:109390, URL http://dx.doi.org/10.1016/j.rser.2019.109390.

Li M, Jiang Y, and Coimbra CF (2017), On the determination of atmospheric longwave irradiance under all-sky conditions, *Solar Energy*, 144:40–48, URL http://dx.doi.org/10.1016/j.s olener.2017.01.006.

Lomanowski BA and Wright JL (2012), The complex fenestration construction: a practical approach for modelling windows with shading devices in ESP-r, *Journal of Building Performance Simulation*, 5(3):185–198, URL http://dx.doi.org/10.10 80/19401493.2011.552735.

Lomas K, Eppel H, Martin C, and Bloomfield D (1997), Empirical validation of building energy simulation programs, *Energy and Buildings*, 26:253–275, URL http://dx.doi.org/10.1016/S 0378-7788(97)00007-8.

Maestre I, Pérez-Lombard L, Foncubierta J, and Cubillas P (2013), Improving direct solar shading calculations within building energy simulation tools, *Journal of Building Performance Simulation*, 6(6):437–448, URL http://dx.doi.org/10.1080/194 01493.2012.745609.

Mahdavi A and Tahmasebi F (2019), *Building Performance Simulation for Design and Operation*, chapter People in building performance simulation, pages 117–145, Routledge, Abingdon, UK, 2nd edition, URL https://www.routledge.com/Buil ding-Performance-Simulation-for-Design-and-Operati on/Hensen-Lamberts/p/book/9781138392199.

Martin M and Berdahl P (1984), Characteristics of infrared sky radiation in the United States, *Solar Energy*, 33(3):321–336, URL http://dx.doi.org/10.1016/0038-092X(84)90162-2.

Mazzarella L and Pasini M (2015), CTF vs FD based numerical methods: Accuracy, stability and computational time's comparison, *Energy Procedia*, 78:2620–2625, URL http://dx.doi.org/10.1016/j.egypro.2015.11.324, 6th International Building Physics Conference, IBPC 2015.

Megri AC and Haghighat F (2007), Zonal modeling for simulating indoor environment of buildings: Review, recent developments, and applications, *HVAC&R Research*, 13(6):887–905, URL http://dx.doi.org/10.1080/10789669.2007.10391461.

Mirsadeghi M, Cóstola D, Blocken B, and Hensen J (2013), Review of external convective heat transfer coefficient models in building energy simulation programs: Implementation and uncertainty, *Applied Thermal Engineering*, 56(1):134–151, URL http://dx.doi.org/10.1016/j.applthermaleng.2013.03.003.

Montazeri H and Blocken B (2017), New generalized expressions for forced convective heat transfer coefficients at building facades and roofs, *Building and Environment*, 119:153–168, URL http://dx.doi.org/10.1016/j.buildenv.2017.04.012.

Montazeri H and Blocken B (2018), Extension of generalized forced convective heat transfer coefficient expressions for isolated buildings taking into account oblique wind directions, *Building and Environment*, 140:194–208, URL http://dx.doi.org/10.1016/j.buildenv.2018.05.027.

NCM (2015), National calculation methodology modelling guide for non-domestic buildings in Scotland, Technical report, Building Research Establishment, Watford, UK.

NECB (2017), National energy code of Canada for buildings, Technical report, Canadian Commission on Building and

Fire Codes, National Research Council of Canada, Ottawa, Canada.

O'Brien W, Gunay B, Tahmasebi F, and Mahdavi A (2017), Special issue on the fundamentals of occupant behaviour research, *Journal of Building Performance Simulation*, 10(5–6):439–443, URL http://dx.doi.org/10.1080/19401493.2017.13830 25.

Peeters L, Beausoleil-Morrison I, and Novoselac A (2011), Internal convective heat transfer modeling: Critical review and discussion of experimentally derived correlations, *Energy and Buildings*, 43(9):2227–2239, URL http://dx.doi.org/10.1016/j.enbuild.2011.05.002.

Perez R, Ineichen P, Seals R, Michalsky J, and Stewart R (1990), Modeling daylight availability and irradiance components from direct and global irradiance, *Solar Energy*, 44(5):271–289, URL http://dx.doi.org/10.1016/0038-092X(90)90055 -H.

Pernigotto G, Prada A, and Gasparella A (2019), Extreme reference years for building energy performance simulation, *Journal of Building Performance Simulation*, 13(2):152–166, URL ht tp://dx.doi.org/10.1080/19401493.2019.1585477.

Pipes LA (1957), Matrix analysis of heat transfer problems, *Journal of the Franklin Institute*, 263(3):195–206, URL http://dx.doi.org/10.1016/0016-0032(57)90927-4.

Prada A, Cappelletti F, Baggio P, and Gasparella A (2014), On the effect of material uncertainties in envelope heat transfer simulations, *Energy and Buildings*, 71:53–60, URL http://dx.doi.org/10.1016/j.enbuild.2013.11.083.

Ren Z, Motlagh O, and Chen D (2020), A correlation-based model for building ground-coupled heat loss calculation using artificial neural network techniques, *Journal of Building Performance Simulation*, 13(1):48–58, URL http://dx.doi.org/10.1080/19401493.2019.1690581.

Sherman M and Grimsrud D (1980), Infiltration-pressurization correlation: simplified physical modeling, *ASHRAE Transactions*, 86(2):778–807, URL https://escholarship.org/content/qt8wd4n2f7/qt8wd4n2f7.pdf?t=lzj6en.

Sowell E and Hittle D (1995), Evolution of building energy simulation methodology, *ASHRAE Transactions*, 101(1):850–855, URL https://www.techstreet.com/standards/ch-95-09-2-evolution-of-building-energy-simulation-methodology?product_id=1716182.

Sparrow EM, Ramsey JW, and Mass EA (1979), Effect of finite width on heat transfer and fluid flow about an inclined rectangular plate, *Journal of Heat Transfer*, 101(2):199–204, URL http://dx.doi.org/10.1115/1.3450946.

Srebric J (2019), *Building Performance Simulation for Design and Operation*, chapter Ventilation performance prediction, pages 76–116, Routledge, Abingdon, UK, 2nd edition, URL https://www.routledge.com/Building-Performance-Simulation-for-Design-and-Operation/Hensen-Lamberts/p/book/9781138392199.

Staley DO and Jurica GM (1972), Effective atmospheric emissivity under clear skies, *Journal of Applied Meteorology (1962-1982)*, 11(2):349–356, URL https://doi.org/10.1175/1520-0450(1972)011%3C0349:EAEUCS%3E2.0.CO;2.

Stephenson D and Mitalas G (1967), Cooling load calculations by thermal response factor method, *ASHRAE Transactions*, 73(1):2018–2024, URL https://nrc-publications.canada.ca/eng/view/accepted/?id=0a164630-2bf0-42b9-a312-695f85ec8a49.

Stephenson D and Mitalas G (1971), Calculation of heat conduction transfer functions for multi-layer slabs, *ASHRAE Transactions*, 77(2):117–126, URL https://nrc-publications.canada.ca/eng/view/accepted/?id=1aca4130-1c7d-419f-a257-fb30f40e67eb.

Strachan P, Svehla K, Heusler I, and Kersken M (2016), Whole model empirical validation on a full-scale building, *Journal of Building Performance Simulation*, 9(4):331–350, URL http://dx.doi.org/10.1080/19401493.2015.1064480.

Tabares-Velasco PC and Griffith B (2012), Diagnostic test cases for verifying surface heat transfer algorithms and boundary conditions in building energy simulation programs, *Journal of Building Performance Simulation*, 5(5):329–346, URL http://dx.doi.org/10.1080/19401493.2011.595501.

Trčka M and Hensen JL (2010), Overview of HVAC system simulation, *Automation in Construction*, 19(2):93–99, URL http://dx.doi.org/10.1016/j.autcon.2009.11.019.

Underwood CP and Yik FW (2004), *Modelling methods for energy in buildings*, Blackwell, Oxford, UK, URL http://dx.doi.org/10.1002/9780470758533.

Walker IS and Wilson DJ (1998), Field validation of algebraic equations for stack and wind driven air infiltration calculations, *HVAC&R Research*, 4(2):119–139, URL http://dx.doi.org/10.1080/10789669.1998.10391395.

Walton GN (1983), Thermal analysis research program reference manual, Technical report, US National Bureau of Standards, Springfield, USA, URL https://www.govinfo.gov/content/pkg/GOVPUB-C13-6176908b08a357a0ac91a8ab3db55b97/pdf/GOVPUB-C13-6176908b08a357a0ac91a8ab3db55b97.pdf, NBSIR 83-2655.

Walton GN (2002), Calculation of obstructed view factors by adaptive integration, Technical report, National Institute of Standard and Technology, Gaithersburg, USA, URL https://nvlpubs.nist.gov/nistpubs/Legacy/IR/nistir6925.pdf, NISTIR 6925.

Wang W, Beausoleil-Morrison I, and Reardon J (2009), Evaluation of the Alberta air infiltration model using measurements and inter-model comparisons, *Building and Environment*,

44(2):309–318, URL http://dx.doi.org/10.1016/j.bui ldenv.2008.03.005.

Wetter M (1999), Simulation model: finned water-to-air coil without condensation, Technical Report LBNL-42355, Lawrence Berkeley National Laboratory, Berkeley, USA, URL https: //simulationresearch.lbl.gov/dirpubs/42355.pdf.

Wetter M (2009), Modelica-based modelling and simulation to support research and development in building energy and control systems, *Journal of Building Performance Simulation*, 2(2):143–161, URL http://dx.doi.org/10.1080/1940149 0902818259.

Wright J (2019), *Building Performance Simulation for Design and Operation*, chapter HVAC systems performance prediction, pages 503–533, Routledge, Abingdon, UK, 2nd edition, URL https://www.routledge.com/Building-Performance-Sim ulation-for-Design-and-Operation/Hensen-Lamberts/p /book/9781138392199.

Yang D (2016), Solar radiation on inclined surfaces: Corrections and benchmarks, *Solar Energy*, 136:288–302, URL http:// dx.doi.org/10.1016/j.solener.2016.06.062.

Yazdanian M and Klems J (1994), Measurement of the exterior convective film coefficient for windows in low-rise buildings, 100(1):1087–1096, URL https://eta-publications.lbl .gov/sites/default/files/34717.pdf.

Zhang K, McDowell TP, and Kummert M (2017), Sky temperature estimation and measurement for longwave radiation calcula- tion, in *Proceedings of Building Simulation 2017*, International Building Performance Simulation Association, San Francisco, USA, URL http://www.ibpsa.org/proceedings/BS2017/B S2017_569.pdf.

Zhou X, Hong T, and Yan D (2014), Comparison of HVAC system modeling in EnergyPlus, DeST and DOE-2.1E, *Building Simu-*

lation, 7(1):21–33, URL http://dx.doi.org/10.1007/s1227 3-013-0150-7.

Index

Note: Information in figures and tables is indicated by page numbers in *italics* and **bold**.

Printed in the United States
By Bookmasters